Anthropogenic Geomorphology

József Szabó · Lóránt Dávid · Dénes Lóczy
Editors

Anthropogenic Geomorphology

A Guide to Man-Made Landforms

Springer

Finally, the possibility that the buildup of greenhouse gases in the atmosphere may cause enhanced global warming in coming decades has many implications for anthropogenic geomorphology.

This valuable book provides an overview of impacts from most types of human activity, demonstrates the value of a historical approach, and although it has a special emphasis on Hungarian research, provides examples from all over the world.

Oxford University, UK Andrew Goudie

Acknowledgements

The authors and editors wish to express their gratitude to Professors Béla Kleb (Budapest) and György Hahn (Miskolc), who read and corrected the Hungarian version of this volume, for their constructive comments. The financial support received from the Hungarian Office for Research and Technology and Pro Renovanda Cultura Hungariae Foundation is also gratefully acknowledged.

Contents

Contributors

Zoltán Baros Institute of Agroinformatics and Rural Development, Károly Róbert College, Mátrai út 36, 3200 Gyöngyös, Hungary, zbaros@karolyrobert.hu

Péter Csima Department of Landscape Protection and Reclamation, Corvinus University of Budapest, Villányi út 35–43, 1118 Budapest, Hungary, peter.csima@uni-corvinus.hu

Péter Csorba Department of Landscape Protection and Environmental Geography, University of Debrecen, Egyetem tér 1, 4032 Debrecen, Hungary, csorbap@delfin.unideb.hu

Lóránt Dávid Department of Tourism and Regional Development, Károly Róbert College, Mátrai út 36, 3200 Gyöngyös, Hungary, davidlo@karolyrobert.hu

Zoltán Ilyés Department for Cultural and Visual Anthropology, University of Miskolc, 3515 Miskolc-Egyetemváros, Hungary, ilyesz@freemail.hu

Zoltán Karancsi Department of Geography and Ecotourism, Institute of Applied Science, Faculty of Education, University of Szeged, Boldogasszony sgt. 6, 6725 Szeged, Hungary, karancsi@jgypk.u-szeged.hu

Attila Kerényi Department of Landscape Protection and Environmental Geography, University of Debrecen, Egyetem tér 1, 4032 Debrecen, Hungary, kerenyi@delfin.klte.hu

Zsuzsanna Lontai-Szilágyi Tourism Department, Ministry of Local Government, József A. u. 2–4, 1051 Budapest, Hungary, szilagyi.zsuzsanna@otm.gov.hu

Dénes Lóczy Department of Environmental Geography and Landscape Conservation, Institute of Environmental Sciences, University of Pécs, Ifjúság útja 6, 7624 Pécs, Hungary, loczyd@gamma.ttk.pte.hu

József Lóki Department of Physical Geography and Geoinformatics, University of Debrecen, Egyetem tér 1, 4032 Debrecen, Hungary, jloki@delfin.klte.hu

Balázs Nagy Department of Physical Geography, Eötvös Loránd University, Pázmány Péter sétány 1/c, 1117 Budapest, Hungary, balazs@afoldgomb.hu

Péter Rózsa Department of Mineralogy and Geology, University of Debrecen, Egyetem tér 1, 4032 Debrecen, Hungary, rozsap@puma.unideb.hu

László Sütő Department of Tourism and Geography, College of Nyíregyháza, Sóstói út. 31/b, 4401 Nyíregyháza, Hungary, sutolaci@nyf.hu

József Szabó Department of Physical Geography and Geoinformatics, University of Debrecen, Egyetem tér 1, 4010 Debrecen, Hungary, wagner@puma.unideb.hu

Csaba Tóth Department of Physical Geography and Geoinformatics, University of Debrecen, Egyetem tér 1, 4032 Debrecen, Hungary, tothcsa@delfin.unideb.hu

Part I
Introduction

Chapter 1
Anthropogenic Geomorphology: Subject and System

József Szabó

Abstract Today the human agent is equal in importance to other geomorphic factors. Although the energy released by human society is insignificant compared to the endogenic forces of the Earth (tectonic movements, volcanic activity, earthquakes), human impact is not only commeasurable to the influence of exogenic processes but even surpasses their efficiency. Exponential population increase involved higher demands and the energy made available to meet the demands resulted in large-scale reworking of surface materials – at an even more rapidly growing rate, a process which is likely to be continued in the future. The subject of anthropogenic geomorphology is the description of the wide and ever-widening range of surface landforms, extremely diverse in origin and in purpose, created by the operation of human society. In a broader sense, artificially created landforms have manifold influences on the environment (e.g. alterations in meso- and microclimate, biota, etc.) and modify natural processes.

Keywords Anthropogenic geomorphology · Subject · System · Classification

1.1 Subject

Chapter presents – in a theoretical approach but through a wide range of examples – the complexity of direct and indirect interactions between the elements of natural systems as well as the ever-intensifying and diversifying external human interventions into such systems. The physical environment of humankind (the envelopes of the Earth, the geographical sphere) is virtually in no part exempt from some kind of human influence, usually cascading through the system and acting back on human society itself. Therefore, it is a logical research objective to

J. Szabó (✉)
Department of Physical Geography and Geoinformatics, University of Debrecen, Egyetem tér 1, 4010 Debrecen, Hungary
e-mail: wagner@puma.unideb.hu

J. Szabó et al. (eds.), *Anthropogenic Geomorphology*,
DOI 10.1007/978-90-481-3058-0_1,

analyse the problems resulting from the above interactions in the most comprehensive approach possible. Now the individual disciplines of earth sciences, specialized on different spheres of the global environment, should devote more attention to research of that kind.

Tasks are also identified in the discipline of geomorphology, which studies landforms, their changes and impacts on other spheres of the global environment:

- Firstly, today the human agent is equal in importance to other factors in the shaping of the Earth's landforms. Although the intensity of its influence depends on the energy released by human society, which is insignificant compared to the endogenic forces of the Earth (tectonic movements, volcanic activity, earthquakes), it is not only commeasurable to the energy which drives most of the exogenic processes but even surpasses the effectiveness of some of them. In the estimation of R.L. Sherlock (1922), the material mobilized by human society on the territory of Great Britain amounted to ca. 30.5 km^3, which would cover the island in ca 13.3 cm depth as opposed to the ca. 7-cm deep layer removed from the surface by exogenic processes over the timespan of 2,000 years. In 1982, Holdgate estimated the annually reworked total soil and rock masses on the Earth at 3×10^{12} t (Holdgate 1982). It is two orders of magnitude higher than the annual total discharge of all rivers (2.4×10^{10} t – according to Judson 1968).
- Secondly, geomorphologists have to study this problem since the geomorphic impact of humans is growing exponentially. Exponential population increase involved higher demands and the energy made available to meet the demands resulted in large-scale reworking of surface materials – at an even more rapidly growing rate. This tendency – at least in the near future – will obviously continue. Consequently, the identification and assessment of impacts will be increasingly important.
- Thirdly, human impact on the Earth's surface does not only influence other natural systems but have a reaction on itself as well to an ever-increasing degree. The rightful judgement that humans are the inhabitants (or sometimes victims) of an environment created (or modified) by themselves is also true for the geomorphic action of humans.

The above considerations clearly pave the way to the formulation of the subject of anthropogenic geomorphology.

In the first approximation, the subject of anthropogenic geomorphology is the description of the wide and ever-widening range of surface landforms, extremely diverse in origin and in purpose, created by the operation of human society. In the above interpretation it is part of dynamic geomorphology since by now human action has established itself as one of the geomorphic agents. If the investigation of the geomorphic action of rivers and its products are labelled 'fluvial geomorphology', an analogous definition can apply to 'anthropogenic geomorphology' too. At a closer look, however, it is found that anthropogenic geomorphology can also be interpreted more narrowly and widely but certainly in a more complex way.

- In a narrower sense, although all human constructions (buildings, industrial plants) modify the appearance of the landscape, they are not regarded as subjects of geomorphological investigation. Such artificial constructions contrast with their environs in size or other properties and undoubtedly influence them. The skyscrapers in Manhattan have fundamentally transformed the landscape but they are outside the scope of geomorphology. In contrast, the Hisarlik Hills over the ruins of Troy and the tells in the Middle East are true geomorphological objects.
- At the same time, the subject of anthropogenic geomorphology is broadened by the fact that the artificially created landforms have manifold influences on the environment (e.g. alterations in meso- and microclimate, biota, etc.). In addition, they may also modify natural processes. New geomorphic processes may be initiated or active processes may be intensified or weakened or even inhibited. As a consequence, new landforms may be generated – not directly by human activities but they would have not formed or not formed in the manner they did without previous human interference. Human geomorphic action may induce cascading environmental changes, whose study obviously lies within the scope of anthropogenic geomorphology. The investigation of the impacts also covers the geomorphic processes induced by the objects which were excluded from anthropogenic geomorphological research above.
- The inevitable complexity of anthropogenic geomorphology derives from the character of natural systems and human activities. Humans interfere with these systems, including geomorphological ones, from outside and thus necessarily disturb the natural order (dynamic equilibrium) of the processes, which has evolved over timespans of various lengths. Man-made landforms are alien to the landscape and through establishing new geomorphological conditions, humans drastically upset the equilibrium. With the appearance of such landforms – if they are not used any more and not maintained by humans – tendencies towards a new equilibrium begin to show. For the society it means uncertainty or occasionally even a threat. On the one hand, it is not easy to predict either the direction of transformation or the nature of the new equilibrium. Both may be deleterious not only for society but also for other natural systems. On the other hand, in the first period of the relaxation time necessary to reach a new equilibrium, changes are rather rapid and may even lead to disastrous consequences. It is far from being a matter of chance that major accidents in the environs of human constructions (dams, waste tips, etc.) usually occur shortly after they were built.

The scope of anthropogenic geomorphology does include not only the study of man-made landforms but also the investigation of man-induced surface changes, the prediction of corollaries of upset natural equilibria as well as the formulation of proposals in order to preclude harmful impacts. The above topics and tasks make anthropogenic geomorphology a discipline of applied character. Its achievements should also serve – in addition to promoting the implementation of socio-economic tasks – environmental protection and nature conservation. When talking about the scope and tasks of anthropogenic geomorphology, another consideration cannot be neglected either. Since most human constructions are located in an environment where natural processes are also active and occasionally even present a hazard to human constructions, human society logically attempts to defend itself against them. In this effort it tries to block or deter natural geomorphic processes and this is how it contributes to geomorphic evolution. Therefore, the protective actions against natural hazards may have implications for anthropogenic geomorphology. If the view is accepted that some structures (like settlements) are so perfectly fitted in the landscape that they function as its natural components, the protection of such structures can be approved even from the viewpoint of nature conservation. The mode of protection, however, is also of importance. When modifying the surface in order to protect us against harmful influences, the considerations of environmental protection and nature conservation have to be taken into account and anthropogenic geomorphological research should cover this field, too.

1.2 System

The thematic complexity and multiple tasks of anthropogenic geomorphology call for a clear internal systemization of the discipline. It is a widespread approach to systemize on the basis whether human action is of *direct* or *indirect* impact on the surface. The direct impact is usually intentional and conscious, leading to clearly recognizable consequences. The less readily identifiable outcomes of indirect human impact, however, are also within the scope of anthropogenic geomorphology and should also be included in its system. Spencer and Hale (1961) classified human actions according to the way their products are related to the initial surface. On this basis they distinguished constructive, excavational, hydrological and agricultural interventions. The latter two can be regarded as the planation of the surface. Similar consideration provides the background to Haigh's (1978) classification. In a simplified form his system is based on the following distinctions:

1. Direct anthropogenic processes
   1.1. constructive
   1.2. excavational
   1.3. hydrological
2. Indirect anthropogenic processes
   2.1. acceleration of erosion and sedimentation
   2.2. subsidence
   2.3. slope failure
   2.4. triggering earthquakes

The system outlined above can be made more logical and complete if, as a first step, human influences are classified from the viewpoint usual for natural geomorphological processes (Table 1.1). In the classification ideas published in the papers

Table 1.1 Geomorphic impacts of human society (with examples) (by Szabó in Szabó J & Dávid L (eds) (2006))

Type of intervention	Land-form type	Direct		Indirect	
		Primary	Secondary	Qualitative	Quantitative
Montanogenic	E	–	Open-cast pits	Subsidences	Fluvial landforms caused by mine water inflow
	P	–	Waste-filled valleys	Accumulation in pits	
	A	–	Waste tips	Bulges around tips	
Industrogenic	E	Cooling lake basins	Quarries for planation	Mass movements on industrial raw material deposition sites	Accelerated erosion by sewage inflow
	P	'Industrial estates'	Slurry reservoirs		
	A	Sockles for windmills	Slag deposition sites		
Urbanogenic	E	Cave dwellings	Loam pits	Cellar collapses	Erosion by runoff from sealed surfaces
	P	P for construction	Garbage disposal sites		
	A	Tells, burial hills	Debris hills		
Traffic	E	Road cuts	Hollow roads	Slumps on embankments	Increased piping
	P	Airfields	Mounds removed		
	A	Embankment	Roadside A		A in culverts
Water management	E	Artificial channels	Navvy pits	Abrasion due to impoundment	Rapid incision
	P	Polders	Cutoffs		
	A	Levees	A by dredging channels		A behind dams
Agrogenic	E	Waterholes	Excavation pits	Rapid gullying	Deflation forms
	P	Terraces	Pseudoterraces	Sheetflow	Silt spreading
	A	Lynchets	Stone ridges	Alluvial fans	Delta expansion
Warfare	E	Moats	Bomb craters	Avalanches caused by explosions	Erosion modified water-courses for defence purposes
	P	Airfields	Destroying settlements		
	A	Earthworks	'Trümmelberge'		
Tourism, sports	E	Recreation lake basins	Field sports (moto-cross) landscapes	Abrasion along recreation lake shores	Accelerated erosion along hiking paths
	P	Sports tracks			
	A	Ski-jumping ramps			

E = excavation processes/landforms; P = planation processes/planated landforms; A = accumulation processes/landforms of Goudie (2007), Erdősi (1987) and Szabó (1993) are also incorporated. Erosion, i.e. processes mostly leading to material deficit, 'negative landforms' (depressions on the surface) finds its counterpart as *excavation* in anthropogenic geomorphology. Accumulation on the surface, mostly producing 'positive landforms' (elevations), can be called constructive, *aggradational* or even here accumulational landforms. The third type of landforms, frequently produced by human action, cannot be referred unambiguously into the categories of natural geomorphic processes: it is called *planation*, which can result from both erosional and accumulational processes under natural conditions. It is often the case in anthropogenic geomorphology, too. Through planation, humans can even destroy landforms created by themselves or by nature (e.g. filling a valley with debris, smoothing a sand dune or even a settlement). In a general formulation, mostly the slope of the surface is reduced. This long-term activity is a particular hazard for the natural environment. *The above listed fundamental types of human intervention into geomorphic evolution can be distinguished within both direct and indirect impacts and it is to the purpose to make this distinction in systemization.* Another aspect of the classification of direct impacts is whether the generation of the landform is the explicit objective of human action or just a more or less unavoidable by-product. When terraces are created on the slopes of hills and mountains for agricultural purposes, the changes in the character of slopes are implemented in the interest of production and, thus, terraces are *primary* landforms here. In another situation, waste tips are accumulated during mining. In this case the 'useless' material has to be deposited in order to extract useful material. Judging from the perspective of the goal of the activity, waste tips are *secondary* landforms here.

Indirect impacts can also be further subdivided. One opportunity is offered by the above section on the subject of anthropogenic geomorphology. One of the large groups of indirect impact includes processes and landforms which would not have been triggered or originated without human action. To cite an example from the field already tackled: the gorges or 'barrancos' on the slopes of waste tips, sometimes of valley size, the alluvial fans at the footslopes of tips or the landslides presenting serious hazard all fall into this category. The processes themselves (e.g. landslides) take place entirely according to physical rules and, as a consequence, the resulting landforms are not at all different from those formed as part of natural systems. Without information on their origin, however, their environmental and geomorphological significance cannot be determined. As processes and landforms of new quality are added to the landscape, they are labelled *qualitative* and are mentioned by Erdősi (1987) as *semi-anthropogenic* processes.

There is another way of operation for indirect human impacts. The activity or the resulting landforms do not induce new processes but only modify the extent and rate of already operating processes together with their consequences. Since no new process occurs here, the impact is not qualitative but quantitative on the natural evolution of an area (*quantitative* changes or natural-anthropogenic processes as called by Erdősi 1987). A good demonstration is erosion by surface runoff. It is well known that forest clearance usually increases runoff and causes floods along rivers. In addition, in the upper section of the catchment valley, incision may accelerate, while increased sediment load of rivers arriving onto the plain enhances the intensity of accumulation there. A change on a more modest scale – although occasionally rather spectacular and rapid – can be, for instance, mine water extracted from depth during mine operation and conducted into surface water-courses. The additional discharge accelerates incision, terraces may take shape overnight, while in other places increased accumulation alters the morphology of the environs of the water-course, thereby causing a transformation also affecting human society.

An obvious aspect is to classify anthropogenic impacts *according to the character of the human activity*. Here another general principle of dynamic geomorphology is applied. Since landforms are usually produced by an interplay of different processes, it is not always easy to distinguish the contribution of the individual processes in the resulting landscape. The whole morphology, however, cannot be interpreted without precise knowledge of the characteristic mechanism and geomorphic impact of the individual processes. To this end, the geomorphic action of rivers and of wind is treated separately as fluvial or eolian geomorphology. As an analogy the impact of the individual branches of the productive activity of human society has to be also investigated separately. Given the large number of the branches, the classification may be too complicated. In a hierarchical solution the main types of social activity are first identified and then further subdivided. In the papers written over recent decades the following fields of anthropogenic geomorphology have been identified:

- *Mining*. The processes involved and the resulting landforms are usually called *montanogenic*.
- *Industrial* impact is reflected in *industrogenic* landforms.
- *Settlement* (urban) expansion exerts a major influence on the landscape over ever-increasing areas. The impacts are called *urbanogenic*.
- *Traffic* also has rather characteristic impacts on the surface.
- As the first civilizations developed, highly advanced farming relied on rivers, *water management* (river channelization, drainage) occupies a special position in anthropogenic geomorphology.
- *Agriculture* is another social activity causing changes on the surface. *Agrogenic* impacts also include transformation due to forestry.
- Although *warfare* is not a productive activity it has long-established surface impacts.
- In contrast, the impacts of *tourism and sports* activities are rather new fields of study in anthropogenic geomorphology.

References

Erdősi F (1987) A társadalom hatása a felszínre, a vizekre és az éghajlatra a Mecsek tágabb környezetében (Impact of Society on the Surface, Water and Climate in the Broader Environs of Mecsek Mountains). Akadémiai Kiadó, Budapest

Goudie A (2007) The Human Impact on the Environment. 6th edn. Blackwell, Oxford

Haigh MJ (1978) Evolution of Slopes on Artificial Landforms. University of Chicago, Blainarch, UK. Dept Geol Res Papers 183

Holdgate MW (ed.) (1982) The World Environment 1972–1982. Tycooly, Dublin

Judson S (1968) Erosion rates near Rome, Italy. Science 160: 1444–1445

Sherlock RL (1922) Man as Geological Agent. Witherby, London

Spencer JE, Hale GA (1961) The origin, nature and distribution of agricultural terracing. Pacific Viewpoint 2: 1–40

Szabó J (1993) A társadalom hatása a földfelszínre (Social impact on the Earth's surface). In: Borsy Z (ed.), Általános természetföldrajz (Physical Geography). Nemzeti Tankönyvkiadó, Budapest, 500–518

Szabó J, Dávid L (eds.) (2006) Antropogén geomorfológia (Anthropogenic Geopmorphology) University notes. Kossuth Egyetemi Kiadó, Debrecen (the Hungarian version of this book)

Part II
Anthropogenic Geomorphology and Related Disciplines

Chapter 2
Human Impact in a Systems Approach

Attila Kerényi

Abstract All the material and energy flows evolved independently from and having existed prior to human presence are considered to be natural in origin. The cycles which are the least influenced by humankind and, therefore, in a 'quasi-natural' state are also in a dynamic equilibrium. The general model of material and energy flows is a result of a generalisation to the greatest extent and reflects the most relevant features of the so-called global geochemical cycles. Geographical factors relevant in geomorphologic processes can be recognised and interpreted in this model. Human society, over its history of approximately 10,000 years, has been intervening into natural processes more and more actively and effectively. Society can influence any of the exogenic geomorphic processes and human impacts can be present at any stage of such processes. There is not a single element of the natural system that would not be influenced by human intervention sooner or later. The degree of changes depends on the intensity of human intervention as well as on the susceptibility of the physical system. Anthropogenic activities without a direct impact on geomorphologic processes can also have consequences on the surface. In their investigation, a system-approach analysis can be of help.

Keywords System approach · Physical system · Geomorphic cycles · Anthropogenic activities

2.1 Some Characteristics of Physical Systems

The physical environment consists of an uncountable number of 'elementary units'. It is a matter of approach as to what is considered to be an elementary unit of a morphological object: an atom, a living being or a rock or soil type, etc.

A. Kerényi (✉)
Department of Landscape Protection and Environmental Geography, University of Debrecen, Egyetem tér 1, 4032 Debrecen, Hungary
e-mail: kerenyi@delfin.klte.hu

J. Szabó et al. (eds.), *Anthropogenic Geomorphology*,
DOI 10.1007/978-90-481-3058-0_2, © Springer Science+Business Media B.V. 2010

Interrelationships between objects (elements) can be of various intensities, direct or indirect. The elements with more direct interrelationships in a given space constitute natural units, *systems*. For scientific research purposes, system elements are detached from other elements of the surrounding world, so that their features can be studied in detail.

As research cannot extend to all subjects, substances and processes, for scientific or practical analysis purposes, a part of reality is isolated from its environment – in most cases only in theory. *Within the system* isolated, *there are functional and structural relationships among system elements*. It has natural boundaries which are to be observed in theoretical considerations. While studying these systems, their relations to their environments have to be analysed.

- *Isolated systems* are material systems with no material and energy input and output. Such systems can only be created in the laboratory – although it is not easy.
- *Closed systems* show energy input and output, but no material exchange with their environment. (The exchange of energy is, though, possible.) Such systems are rare on Earth.
- Between *open systems* and their environments both energy and material exchanges take place. Energy flow is mainly bound to material flow, i.e. materials drifting among systems carry a certain amount of energy. This can be, e.g., adsorbed heat energy or potential chemical energy. Open systems largely maintain their structures despite the material and energy flows through them.

All natural systems are classified as open systems. The maintenance of their basic structure does not mean system permanence; changes are significant and characteristic. Theoretically, changes within the system can be described accurately in case their initial and final stages are identified. In practice, however, information on the course of changes, in other words, the description of the series of intermediary stages by which the process of change is revealed, may often be relevant.

The persistence of system structure is a result of perpetual material and energy flows. Take a water regime as an example: it sustains all its essential features over the long term, while the water itself exhibits constant motion and is characterised by periodic water input (precipitation) and constant water output (estuary discharge + evaporation). This material flow is also pertained to energy flow. Water of high heat capacity absorbs heat from and emits it to the environment and, at the same time, represents a significant amount of kinetic energy which causes surface erosion, provides the energy coverage of bedload transport and also facilitates the operation of hydroelectric plants. Material and energy flows cause alterations in the system (river bed formation, soil erosion, changes in slope inclination), all essential features of the water regime (the shape of the catchment area, the number of major watercourses, mean estuary discharge, flood return intervals, the rate of evaporation), however, remain stable. This state of the system, though not static, can be regarded as equilibrium. A feature of natural systems is that they mostly tend towards a *dynamic equilibrium* through their functioning.

Natural systems are also connected to each other. Systems closely interrelated comprise a complex system; moreover, such complex systems can further build higher-level systems. These complex and multiple complex systems make up Earth's physical environment.

The uniformity of this system with extremely complex structure and function is explained by material and energy flows creating functional relationships by percolating into sub-systems. For the *material and energy flows which occur in the geospheres, certain cyclicity* is typical and it primarily involves the transport of chemical compounds and elements during which chemical and/or chemico-physical transformations take place. Such cycles are called *geochemical cycles*.

2.2 General Model of Material and Energy Cycles and Its Relevance for Geomorphology

Most of the material and energy flows had operated on Earth well before human presence. For instance, water reached the atmosphere through evaporation, solid particles and various gases as a result of volcanic activity, and later deposited on the soil or bare rock surfaces or in water or adsorbed on soil particles. From here, due to the dissolving effect of precipitation, they could have been carried to the groundwater and surface waters, as well as to seas and oceans through water systems. Plants and animals also contributed to the operation of material and energy cycles (photosynthesis, respiration). All the material and energy flows evolved independently from and having existed prior to human presence are considered to be natural in origin.

All geochemical cycles can be demonstrated by a model for the movement of chemical compounds or elements. The model shows the directions and courses of this movement as well as environmental objects where the given element or compound remained for a longer period of time (*reservoirs*), and all quantitative data available for them.

The number of natural storages (reservoirs) depends on the detail of modelling, i.e. on its resolution. An important feature of the cycle is *flux*, defining the amount of material transported along a given course within a defined period of time (usually a year). When the concentration of an element/compound remains constant in the reservoir, it indicates a balance between the input and output elements/compounds. This refers to a dynamic equilibrium. The cycles which are least influenced by humankind and, therefore, in 'quasi-natural' state are also in a dynamic equilibrium.

Having recognised the system is in such a state, the *residence time* of given elements in given reservoirs are defined as follows:

$$\text{residence time} = \frac{\text{the amount of the given element in the reservoir}}{\text{the input (or output) rate of the element (quantity per annum)}} \qquad (2.1)$$

In case the mass of *sodium* dissolved in the ocean is 15×10^{18} kg and it receives an additional annual amount of 10^{11} kg, the residence time is 150 million years.

As seen from this data, the time periods involved are rather long on the human scale. In the ocean, the residence time of the vast majority of elements is on the scale of million years. There are, however, geochemical cycles with shorter period as e.g. the terrestrial water cycle relevant from the point of view of geomorphology (precipitation reaching the surface, runoff, evaporation, residence in the atmosphere). Here residence times can be measured in the order of days and months.

The general model of material and energy flows is represented in Fig. 2.1. (O'Neill 1985). This model is a result of a generalisation to the greatest extent and reflects the most relevant features of the so-called *global geochemical cycles*. The upper cycle of the figure represents processes taking place on the surface of the lithosphere and in the two resilient geospheres (atmosphere, hydrosphere) whereas the lower one runs in the lithosphere. Water plays the most important role in material transport (precipitation, surface and subsurface waters on land as well as seawater).

Geographical factors relevant in geomorphologic processes can be recognised and interpreted in the figure. Sediment accumulation in the oceans eventually results in the formation of sedimentary rocks, which are transformed, metamorphosed and, in a modified form, take part in the folding of mountains in the course of plate tectonic movements. Eventually, plate tectonic cycles control macro-scale geomorphologic elements such as the orogenic structures, tectonic trenches, depressions and others.

Soil formation is a result of physical and chemical weathering processes among which biogenic weathering has a predominant role. Sheet wash caused by atmospheric precipitation, then stream erosion contributes to the evolution of a range

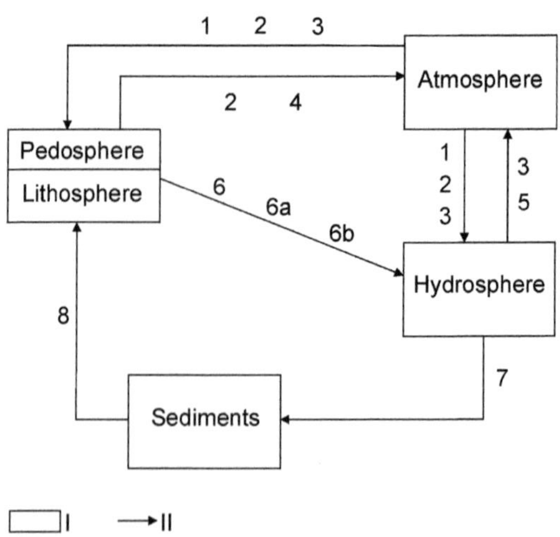

Fig. 2.1 The general model of material and energy cycles (geochemical cycles) Modified after O'Neill (1985) 1. precipitation, 2. dust, 3. vapourised water drops of the sea, 4. gas loss, 5. gas generation, 6. rivers (dissolved and suspended substance), 6a. glaciers, 6b. coastal erosion, 7. sedimentation and sediment deposition, 8. material exposure by folding and tectonism. I. reservoirs, II. material and energy flows

of geomorphological features. In a broader sense, sediment transport by water, and consequently, geomorphic action, that of glaciers as well as wave erosion along coasts and lake shores are also significant processes in the transformation of the surface. In addition to water, wind can also shape soil and rock surfaces (primarily unconsolidated and unvegetated sedimentary rock surfaces) – especially in arid and semi-arid regions.

All these processes are fundamental factors in geomorphic evolution. Some of them – mainly the impact of endogenic forces – cannot be altered by human activities; however, exogenic forces are influenced either directly or indirectly, deliberately or spontaneously by human action.

2.3 The Impact of Human Activities on Geomorphologic Processes

The figure illustrating the major geomorphologic cycles only reflects natural processes and only provides a general overview. Human society, over its history of approximately 10,000 years, has been intervening into natural processes more and more actively and effectively. The establishment of the first settlements built of rocks and/or clay also represented the first major anthropogenic geomorphologic impact, as rocks and clay had to be excavated resulting in the formation of depressions on the surface. During its history, humankind has been carrying out more and more production activities, which directly or indirectly altered the physical environment. In the flowchart below, the role of some anthropogenic environmental impacts regarded to be relevant in modifying geomorphologic processes are examined (Fig. 2.2).

Endogenic forces, responsible for the formation of major landforms, cause isostatic uplift and subsidence of the crust, vertical movements, folding and faulting, as well as by volcanism, producing hills and mountains of volcanic origin. These processes are accompanied by earthquakes. A common feature of these forces is the increase of differences in height measurable on the surface and the formation of slopes of various inclinations. Thus, gravitation has an active role in the displacements of crust material. In erosion, the inertial resistance of rocks also has a part to play (Fig. 2.2).

The impact of *exogenic forces* can be modified by a range of activities even for cases such as solar radiation. By the extinction of natural vegetation (whether it is carried out for agricultural or construction purposes), the amount of incoming solar radiation reaching the soil and/or rock surface changes significantly. This greatly increases the heating and cooling of rocks (temperature extremities of the 'bare' surface are also intensified by the increase in outward radiation) and accelerates physical weathering and, eventually, contributes to increased erosion rates.

Construction activities can also transform the heat input of the surface. Extensive built-up areas (e.g. cities), however, also exert an impact on precipitation. Sealed surfaces have an even stronger influence on surface runoff, a decisive factor of

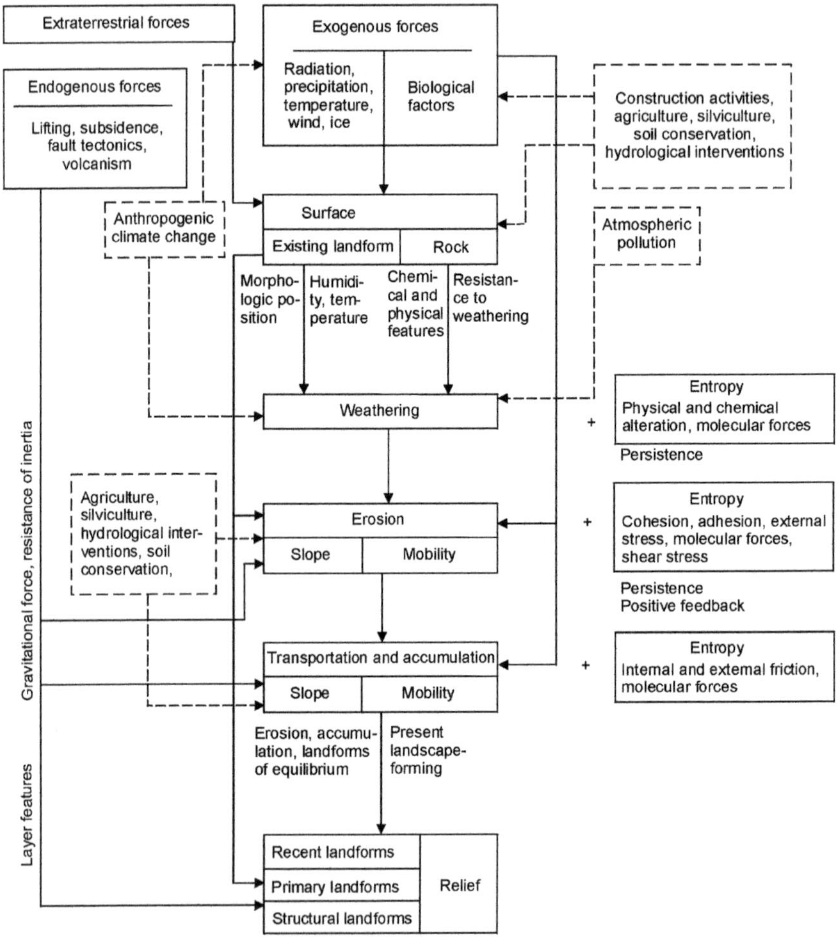

Fig. 2.2 Geomorphologic processes, influencing factors and the predominating entropy. After Bremer (1989), modified by including major anthropogenic impacts

water erosion. Linear structures (roads and motorways, railway lines and others) also significantly modify the geomorphic action of water.

Agricultural and forest management practices also greatly alter surface runoff: extensive clear-cuts may result in rapid erosion as the amount of precipitation retained by vegetation may drop abruptly.

Soil conservation measures (such as terracing), which aim at efficient rainwater retention, result in a decrease of the amount of surface runoff and, thus, have an adverse impact. Levelling and other activities may have direct geomorphological consequences.

Hydrological interventions (cut-offs, construction of dams, etc.) for flood control influence morphologic processes both directly and indirectly: they create new landforms as well as modify the process of channel erosion and even that of sediment accumulation.

The impact of *atmospheric pollution* on erosion is hard to estimate. However when considered that contaminants reaching the atmosphere in a considerable quantity (sulphur dioxide, nitrogen oxides, ammonia) undergo chemical transformations, and, as a result, acidic deposition takes place on the surface, it becomes clear that acidic substances intensify the process of chemical weathering and result in increasing erosion. This is accompanied by an increase in entropy.

Entropy is a thermodynamic parameter of a state indicating the amount of energy the can be transferred from one system to another in the form of work. When the system entropy is zero, its total energy can be applied for work. In case its energy transformable to mechanical work is zero, system entropy reaches its maximum. As seen in Fig. 2.2, geomorphic processes like chemical weathering, erosion, transportation and accumulation are accompanied by an increase in entropy. Here physical, chemical and combined chemico-physical processes operate.

Man-induced climate change impacts geomorphologic processes in a way basically modifying the operation of all exogenic forces. It is obvious that any change in the climate necessarily entails alterations in precipitation (amount, time distribution, intensity) and temperature. If the degree of cloudiness alters, solar radiation also undergoes changes. Obviously the overall air circulation should not remain constant either and thus, wind conditions and, as a consequence, the rate of deflation is also modified. Climate change also has an influence on glacial processes. It is well known that the transgression of glaciers in colder and more humid periods may result in increased ice motion velocities. Examples, however, can be found for the opposite too. Today, most of the glaciers are retreating and this is related to warming climate. It is also evident that human-induced climate change can also significantly alter the physical and chemical weathering of rocks.

Despite all these apparently significant impacts of climate on geomorphic processes, some representatives of anthropogenic geomorphology do not consider human-induced climate change to be an impact of human origin. This is explained by the fact that during the Earth's history, a number of major natural climate changes took place with significant geomorphological consequences, and it is practically impossible to decide to what degree the anthropogenic character of climatic factors can be taken into account in recent geomorphologic processes, whether a landform developed under decisively climatic influence can be regarded anthropogenic in origin.

In addition to the process of natural erosion, the degree of accumulation is greatly modified by interventions of agriculture and forestry, soil conservation and water management measures (Fig. 2.2).

It can be seen that society can influence any of the exogenic geomorphic processes as well as that human impacts can be present at any stages of such processes. Whether landforms developed as an outcome of anthropogenic processes depends on the extent of human impact. This is, in certain cases, rather obvious and easy to define; however requires detailed analysis in others in order to classify given forms. The researchers' work is aggravated by the complicated, 'network-type' interrelationships of natural systems, i.e. an impact on any single physical element induces a number of indirect impacts on other factors. Section 2.4 is a review on some related principles.

2.4 Indirect Human Impacts on Physical Systems

Landforms can be regarded as elements of physical geographical systems. Before the emergence of humans, the factors of physical geographical systems were *topography* (landforms), *water* (surface and underground), *air/climate, rocks, soil and the biota.* In the literature on geography, regarding the nomenclature of these factors, major or minor differences are present according to the various sources (for details, see Kerényi 1995; Csorba 1997; Lóczy 2002). Among these, the most relevant difference is the indication of climate, instead of air, as a physical element. For air, water, organic world, soil and rocks, various substantial qualities are represented. The first two are decisive transporting agents in nature, thus playing a fundamental role in geomorphologic processes as well. Climate is, obviously, more than air and is also different in quality: a state of environment resultant from the interrelationship of the physical geographical factors, mainly described by the atmospheric parameters, e.g. air temperature at 2 m height, wind velocity (i.e. the horizontal component of air flow), relative air humidity and others. In the meantime, solar radiation is also a fundamental factor of climate, which drives the functioning of the whole organic terrestrial system.

For such reasons, using climate instead of air in the physical geographical system model makes the representation of the system more complete. (It should be noted here that renowned landscape researchers also analyse natural systems in the same approach; see Haase 1978; Leser 1991).

Let us investigate the interrelationships among the system elements within the physical geographical systems and alterations in the whole system caused by some characteristic impacts of human activities – with a special focus on topographic changes.

Various geometric forms indicate the decisive substantial quality of the system elements in Fig. 2.3 (Kerényi 2007). Rectangles represent the predominantly inorganic natural elements; a circle indicates the living world and a polygon marks soils, which are primarily interim between organic and inorganic. It is an inevitable fact that a rich biota exists in natural waters, but this is so marginal compared to the mass of the hydrosphere that the predominance of water as an inorganic agent should not be argued. From our point of view, the role of mechanical energy of the flowing water in geomorphic evolution is also taken into account and, in this respect, the amount of aquatic organisms is irrelevant.

Soil is a different case in many respects. This layer is sometimes only 20–30 cm deep and the 'buffer layer' of geomorphic processes is 1.5 m deep in general. While particles of the soil humus layer are eroded from the surface by wind or water, at the lower boundary of the soil section, the inorganic rocks undergo biogenic

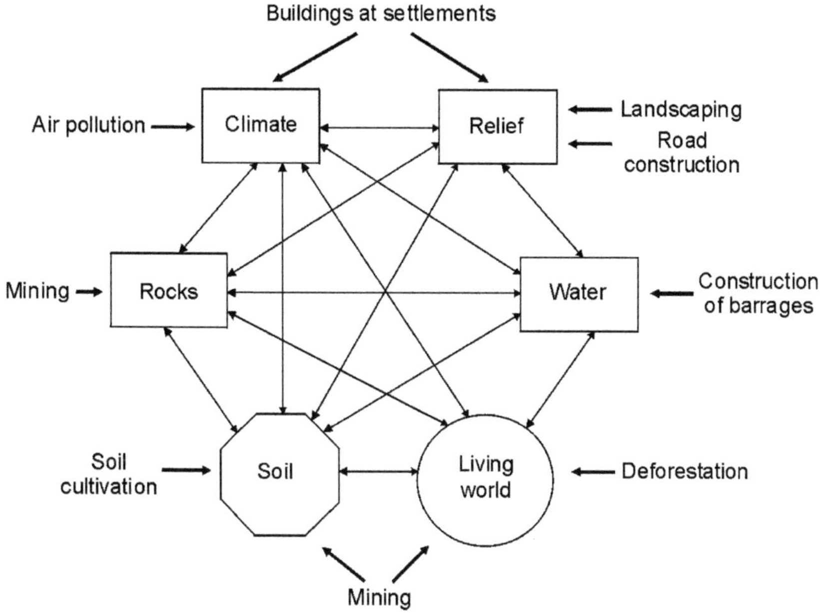

Fig. 2.3 Some typical impact points of the physical geographical system and anthropogenic activities (by Kerényi 2007)

weathering – i.e. soil formation takes place. When these two processes are balanced, the thickness of soil layer is constant, the surface of the given slope has a loss in substance and, depending on whether erosion is along particular lines or uniform over the surface, morphological changes follow.

Soil is transitional between organic and inorganic environmental elements in respect to the fundamental processes taking place within, i.e. humus formation, are determined by its extremely rich biota. Soils are also interwoven by the root system of superior vegetation; vegetation and soils are indispensable to each other: vegetation receives water and nutrients necessary to its sustenance, whereas the removal of vegetation results in soil loss (accelerated soil erosion).

This supports the representation of the interrelation between the *biota* and soils in our system model. It is claimed that this is one of the closest interrelationships in physical geographical systems (Fig. 2.3).

As mentioned above, the impact of the biota on *rocks* (biogenic weathering) is, in many cases, direct. Plants (so-called pioneer plants) able to settle on bare rocks by their root acids and micro-organisms living in symbiosis on their roots, trigger the process of soil formation. Meanwhile, the hardness and chemical composition of rocks defines what species are capable of living on them.

Water is a necessary living condition of the organic world; the vegetation on the surface, however, has a major influence on the water balance of a given geographical region, partly through transpiration and partly through hindering runoff. The presence of water also determines the processes of rock weathering and also plays a part in the water storage of rocks (depending on their structure) as well as defines the amount of runoff. The same interrelationship is observed for soils and water.

Climate, on the one hand, exerts a direct influence on the living conditions of organisms and, by this, on species composition. (Here we refer to stenotopic and eurytopic species.) Undoubtedly, climatic factors strongly influence inorganic environmental elements, e.g. greater variation in the temperature enhances physical weathering; warmer air of low humidity increases evaporation. Climate is also a decisive factor in the formation of *topography*, as falling precipitation and wind are both capable of carrying a significant amount of solid substance resulting in the constant modification of landforms.

The reaction of inorganic environmental elements on the climate can also be represented by recalling well-known phenomena. The colour of the rock surface determines the degree of reflected radiation, influencing the heating of air. Water, by its rather high heat capacity, contributes to balanced climatic conditions; the extension of water surface has an influence on the level of direct evaporation. Diverse topography, the inclination and exposure of slopes modify the value of irradiation at various locations and greatly influence local (micro- and/or meso-) climate.

Though the internal system of relationships of our model has only been roughly represented, it still seems to be rather complicated. We did not aim at describing the interrelationships in detail as it would be the subject of a full course in physical geography. We mainly intended to draw attention to the fact that the elements of physical geographical systems are closely interrelated; thus any impact on one of its elements would never be restricted to the alteration of that very element.

Several *anthropogenic activities*, with either direct or indirect impact on the development of anthropogenic features are represented in Fig. 2.3. Some of them are tackled below and the consequences of the given anthropogenic intervention will be briefly analysed.

Prior to construction, terrain levelling is usually carried out. Depending on the topographical features, it is accompanied by a restrained or a more comprehensive geomorphological transformation, in certain cases even resulting in a direct damage to the natural environment as, for instance, in the Buda Hills within the territory of Budapest, where, as a consequence of urban development, a number of minor caves have been destroyed. Terrain correction inevitably results in the emergence of human-made landforms, among others, terraces or road cuts.

In addition to the direct transformation of the surface, construction works also have other consequences. As a result of building activities, surface runoff is modified locally by changing slope angle or by surface sealing. In the latter case, an extremely wide range of artificial materials of variable permeability are applied. The rate of erosion can be modified to some degree by the altered runoff water amount and its modified speed: it can either increase or cease when, e.g., the whole amount of rainwater runoff is canalised (see also Chapter 3).

Whichever of the above-mentioned interventions occurs, there will be a change in the rate of (evapo)transpiration as well as the reflected radiation depending on the surface cover, i.e. the microclimate is influenced.

The consequence of terrain correction is the destruction of the biota and most of the soil. Although, some kinds of landscape restoration are usually done after construction, this 'new biota' cannot be compared to natural or quasi-natural ecosystems in biodiversity and in their capability for adjustment. In artificial ecosystems, these fundamental features are lacking.

A case study is presented below, where the primary aim was not the modification of topography but the construction of a barrage and associated system channels and dam for the purpose of irrigation. During the construction of the Kisköre Barrage Scheme (on the Tisza River, in the Great Hungarian Plain), it became evident already in the first stage of works that such an investment could not be accomplished without alterations made to the terrain. Terrain correction impacted, in this case, a more significant area compared to an average construction in an urban area. In addition to the barrage, the establishment of flood-control dykes seemed to be necessary. Vast earthworks of the channel system included not only the establishment of linear channel bed forms but the accumulation of the excavated soil. Whereas the primary aim was water supply for irrigation, everything changed in the landscape: an enormous backwater surface increasing evaporation, moderating temperature extremes in the surroundings, raising groundwater levels and changing the living environment around. In the flooded area, terrestrial organisms either perished or migrated elsewhere, soil cover was destroyed whereas organisms living in lotic habitats are replaced by others typical of lacustrine environments. This increased inundation will exert a pressure on rocks, and their repeated subsidence will trigger minor earthquakes. (At areas of higher relief, terrain correction can cause landslides or rockfalls when the 'support' of soil or rock layers was disrupted.)

There is not a single element of the natural system that would not be influenced by human intervention sooner or later.

The conclusion drawn from the facts mentioned above is simple: any human activity represented in Fig. 2.3 necessarily has indirect impacts, altering all elements of the physical system to some extent. The degree of changes depends on the intensity of human intervention as well as on the susceptibility of the physical system. During the planning of activities resulting in direct topographic changes, such indirect impacts are important to be analysed, however, in the other way round: anthropogenic activities without a direct impact on geomorphic processes can also have consequences on the surface. In their exploration, a system-approach analysis is helpful. In the following chapters of this book, the direct and indirect geomorphologic impacts of all human activities mentioned above are studied. Here, it should be emphasised that in order to minimise environmental damage, production activities of the society must be adjusted to the susceptibility of physical systems.

References

Bremer H (1989) Allgemeine Geomorphologie. Gebrüder Borntraeger, Berlin, Stuttgart

Csorba P (1997) Tájökológia (Landscape Ecology). Kossuth Egyetemi Kiadó, Debrecen

Haase G (1978) Zur Ableitung und Kennzeichnung von Naturraumpotentialen. Petermanns Geogr Mitt 122: 113–125

Kerényi A (1995) Általános környezetvédelem (Environmental Protection). Mozaik Oktatási Stúdió, Szeged
Kerényi A (2007) Tájvédelem (Landscape Conservation). Pedellus Tankönyvkiadó, Debrecen
Leser H (1991) Landschaftsökologie. UTB 521. V. E. Ulmer, Stuttgart
Lóczy D (2002) Tájértékelés, földértékelés (Land Evaluation). Dialóg Campus Kiadó, Budapest-Pécs
O'Neill P (1985) Environmental Chemistry. George Allen and Unwin, London

Chapter 3
Anthropogenic Geomorphology in Environmental Management

Dénes Lóczy

Abstract Human activities with geomorphic impact become an integrated part of environmental management, encompassing both the utilization of environmental resources and the simultaneous protection of environmental values. The objectives of anthropogenic geomorphological research in environmental management include the prediction of the course of geomorphic evolution after human intervention over various time-scales (directly providing geomorphological data for environmental management); promoting the restoration of degraded landscapes; the identification of landforms worth preserving (serving landscape conservation through estimating the conservational value and the vulnerability of landforms). The rehabilitation of the landscape is particularly important in industrial and mining regions. River channel and floodplain restoration measures are also in the forefront of interest and serve landscape ecological purposes. Fulfilling these requirements, geomorphology, in the 21st century will acquire an even more important role in environmental management.

Keywords Environmental management · Environmental geomorphology · Geomorphological hazards · Land reclamation

3.1 Introduction

Globally, the extent and rate of man-induced geomorphic processes has become comparable to natural ones and in the densely populated areas of the Earth they even exceed the intensity of naturally occurring physical processes, traditionally studied by geomorphologists. The evolution of the surface is influenced by human society

D. Lóczy (✉)
Department of Environmental Geography and Landscape Conservation, Institute of Environmental Sciences, University of Pécs, Ifjúság útja 6, 7624 Pécs, Hungary
e-mail: loczyd@gamma.ttk.pte.hu

J. Szabó et al. (eds.), *Anthropogenic Geomorphology*,
DOI 10.1007/978-90-481-3058-0_3, © Springer Science+Business Media B.V. 2010

both indirectly (e.g. transforming the thermal budget or the runoff conditions of surfaces through forest clearance) and directly (creating accumulational landforms like waste tips as well as excavational depressions like open-cast mining pits, road and railway cuts and many others). Therefore, the human activities with geomorphic impact become an integrated part of *environmental management*, i.e. the long series of tasks which encompass all stages of the utilization of environmental resources on the one hand and the simultaneous protection of environmental values on the other. It involves the identification of the goal of environmental intervention (e.g. mineral extraction, soil amelioration or any kind of construction), policy formulation, project planning, implementation and policy/project evaluation (Cooke and Doornkamp 1990). Therefore, environmental management is not purely an economic concept. It includes activities like baseline survey, reconnaissance research, planning, development, securing financial and social preconditions and the assessment of predictable environmental impact, technical implementation, environmental monitoring – in the meantime mitigating environmental damage deriving from production or services and continuously developing the scope of the resources available.

3.2 Applied or Environmental Geomorphology

In what aspects is geomorphological research, focusing on geomorphic processes and the landforms they produce, of direct benefit for environmental management? Historically, Gilbert's (1917) study on debris transport induced by hydraulic mining on the Sacramento River, California, is regarded the first geomorphological paper of direct relevance to environmental management. The application of geomorphological knowledge accumulated from more than a century of scientific research allows the optimal coexistence of humankind with geomorphic processes and landforms and optimal adjustment to them.

Coexistence, however, does not mean domination, 'mastering nature'. In the 1960s and 1970s Soviet *'constructive geography'* (Gerasimov 1968) set a radical improvement of environmental conditions as an objective – in order to serve the ambitious social plans dictated by Marxist ideology. (An example is the so-called Davidov Plan, i.e. the transfer of water from the 'useless' northward flowing rivers of Siberia to the drought-stricken lands of Central Asia – Kelly et al. 1983). It is less widely known but similar rainbows were also chased in the United States of America, suffice it to mention an engineering geomorphological publication of the US Geological Survey entitled 'Nature to be Commanded' (Robinson and Spieker 1978). (It is only fair to note, however, that the full citation from Francis Bacon reads as 'Nature to be commanded must be obeyed'). In the 20th century, humankind infinitely overestimated both its capability to change the physical environment and the expected benefits of intervention into nature. Although water transfer projects remain on the agenda in many parts of the world (e.g. in India) and some of them are even in the stage of implementation (like the 'Great Man-Made River' water pipeline in Libya and the south-to-north water transfer canal in China – see Case study 1), there are indications that humans are beginning to show a more humble attitude to the physical environment, including topographic conditions.

In the first decade of the 21st century a more sober concept of applied geomorphology incorporates the following tasks:

- to promote the exploration of natural resources (mostly through the mapping of their spatial distribution);
- to help avoid natural disasters (through the identification of natural hazards, the study of their origin and monitoring their evolution and estimating their extent and frequency);
- to predict the course of geomorphic evolution after human intervention over various time-scales (directly providing geomorphological data for environmental management);
- to identify landforms worth preserving (serving landscape conservation through estimating the conservational value and the vulnerability of landforms);
- to provide background information for the restoration of degraded landscapes (through reconstructing topographic conditions prior to human intervention).

Engineering geomorphology may provide even more direct contributions to urban development, construction of buildings, transportation facilities, utilities, flood-control structures and recreational facilities (Cooke and Doornkamp 1990). The particular benefits of geomorphological knowledge in these tasks can only be illustrated by case studies.

From a purely organizational aspect, geomorphologists can participate in environmental management planning in three various modes (Graf 1996):

- as experts entirely independent from economic or investment companies (mostly university experts);
- as external consultants, who are in contract with partners responsible for the implementation of environment management tasks or
- as experts directly employed by decision makers.

3.3 Resource Exploration and Geomorphology

A task of environmental management is the delimitation of mineral reserves. How can geomorphological information be used in exploration for mineral resources? Data on the conditions and rate of operation of various geomorphic processes (entrainment, transport and deposition) and on their distribution, size, structure and age are gathered by fundamental geomorphological research and represented on geomorphological maps. It is of particular significance here that – under certain conditions – even general-purpose geomorphological maps (without special contents on engineering geomorphology) provide information on the spatial distribution of some mineral reserves. To locate perspective sites where it is worthwhile to start geological investigations (cost-intensive drilling), analogies are analysed (Bennett and Doyle 1997). For the construction of generic exploration models, geomorphological information is particularly useful since the localities, for instance, with raw materials for the building industry (limestones, sandstones) can also be identified on a geological map but the quality of the material may be very heterogeneous within a patch marked by the same colour on the geological map. The map of landforms can supply valuable additional information in this respect – particularly for widespread and relatively low-value resources, which are not economic to transport over longer distances (see Case study 2). Deciding on the utilization of any type of model, search parameters are of great importance in exploration. In the case of other building materials (building stones), the weathered nature of rock bodies can be of decisive importance. The degree of weathering can be estimated from the duration of rock exposure, also deduced from geomorphological maps.

3.4 Geomorphological Hazards

In addition to environmental resources, the physical environment also presents hazards, which endanger the utilization of resources or even threaten human lives. *Geomorphological hazards* constitute a subgroup of environmental hazards (Embleton et al. 1989) and include events and processes which cause a change in the characteristics of the Earth's surface detrimental to human society and its activities. According to the definition accepted by the Working Group on 'Geomorphological Survey and Mapping' of the International Geographical Union, natural processes may induce such changes but for quite a number of them human intervention is – directly or indirectly – responsible and thus the latter are among the research subjects of anthropogenic geomorphology (Table 3.1). Disasters only happen in densely inhabited regions, where significant damage can result. The concept of *vulnerability* expresses the potential damage to human lives and properties, which is increasing with growing population numbers. The consequences of disasters are also becoming ever more complicated. Destructive earthquakes, volcanic eruptions, mass movements, inundations or droughts are particularly tragic events in developing countries, where vulnerability is higher. In order to avoid catastrophes or mitigate their damage, costly research projects are under way. However, Table 3.1 shows that man-induced hazards are not easy to prevent (most of them are only moderately avoidable) but their prediction do not usually present an insolvable task.

Although inducing a volcanic eruption is beyond the capability of humans, several earthquakes – and most of the resulting damage – are attributed to social activities. The filling and the depletion of reservoirs as well as mining activities can equally induce earthquakes (see Case study 2) (Hudyma 2004). Some of the naturally triggered earthquakes involved huge damage because of human factors. The infamous 1906 San Francisco earthquake caused the largest destruction next

Table 3.1 Classification of geomorphological hazards (modified from Embleton et al. 1989)

Main group of process	Potentially damaging process	Rapid/ catastrophic?	Avoidable?	Predictable?
Natural hazards				
Seismic	Earthquake, landslide, vertical dislocations	Yes	No	No or hardly
Volcanic	Eruption, lava flow, lahar, mudflow, ashfall, damming water-courses, volcanic jökulhlaup	Yes	No	Hardly
Glacial	Ice advance, glacier collapse, lake impoundment, jökulhlaup, icy flood	Partly	Hardly	Well or moderately
(Occasionally) Man-induced				
Gravitational	Mass movements (fall, creep, slide, flow)	Generally yes	To some extent	Well or moderately
Subsidence	Sagging, collapse	Rarely	To a small extent	Well
Fluvial	Flood, riverbank and soil erosion	Generally yes	Well or moderately	Well or moderately
Nival	Avalanche	Yes	Moderately	Well or moderately
Coastal	Coastal erosion, sedimentation, saltwater intrusion	No	Well or moderately	Well
Aeolian	Deflation, dune mobility, burial under sand and dust	No	Hardly or moderately	Well

to the Bay, in areas reclaimed from the sea. The unconsolidated sand responded to vibration by liquefaction, which largely reduced its shear strength.

Applied geomorphological research is also considered relevant to risk analysis (Panizza 1996). The description of seismic hazard begins with morphotectonic investigations followed by morphometric analysis and geomorphological mapping (slope inclination, previous landslides). Then the scope of object threatened is surveyed and vulnerability is estimated and mapped in zones on a range from major, through moderate to minimum vulnerability. The zoning underpins limitations in land use (e.g. building licences).

The monitoring of processes is aided by photogrammetry, digital elevation models (DEM) and simulational prediction models. Risk (*R*) is calculated from the probability of hazard (*H*), vulnerability (*V*) and the elements threatened (*E*) (Varnes and IAEG 1984):

$$R = H \times V \times E \tag{3.1}$$

Gravitational movements can also be triggered by human activities. Subsidence commonly accompanies the extraction of drinking and irrigation water and of mineral oil and natural gas; thawing of permafrost and oxidation of peat (both may result from forest clearing); karstification and piping processes; undermining and other influences. Italian geomorphologists have played a leading role in landslide hazard mapping and risk analysis (Guzzetti et al. 1999). Surface instability is a major hazard in Italy, where between 1950 and 2000, 521 landslides demand a toll of 4408 victims, while floods involved 1040 deaths and earthquakes 4160 deaths (Guzzetti 2003). For the combination of landslide and flood, the tragedy of the Vaiont Valley in 1963 is the best known example.

In most countries, however, the largest number of natural disasters is associated with floods. For geomorphologists floods are efficient means of shaping the surface (Gregory 1985) but flood risk analyses – increasingly common also in Hungarian literature (Nagy and Tóth 2001) – have become an integral part of environmental management (Table 3.2 – Hooke 1988).

In addition to the above rapid geomorphological hazards, more gradual geomorphic processes under human impact are also significant in environmental

Table 3.2 Societal problems associated with rivers (modified from Hooke 1988)

Origin of impact	Process	Consequences on the environment and society
Inherent natural	River channel instability and changes, e.g. migration, cut-offs	Destruction of structures, loss of amenity value, navigation difficulties, boundary disputes
	Flooding	Destruction of structures and communications, danger to life and property, destruction of crops, interruption of activities, drainage difficulties
External natural	Climate change (changes in discharge)	Problems in water supply, vegetation change, land use change
Direct human	River regulation, channelization	Damage mitigation but need for permanent maintenance
	Water abstraction	Problems in water distribution
	Waste disposal	Deteriorating ecological conditions, loss of amenity value, decrease in fish stocks
	Irrigation	Soil salinization
	Drainage, especially agricultural	Groundwater alterations, vegetation and land use changes
Indirect human	Land use change (especially deforestation or afforestation)	Accelerated soil erosion, channel sedimentation, increased flood hazard
	Urbanization and roads	Growing extremities in discharge, channel sedimentation
	Mining	Channel sedimentation, deteriorating ecological conditions
	Agricultural practices	Drought, soil salinization

management. More information on the most significant of such processes, soil erosion, is found in Chapters 5 and 7.

3.5 Prediction of Geomorphic Evolution

Environmental policy is not restricted to overcoming crises (handling geomorphological hazards) but also encourages development, directly promoting the solution of some engineering tasks, like pipeline construction in permafrost areas or protecting structures from salt weathering in arid regions (Cooke and Doornkamp 1990). A profound understanding of periglacial and weathering processes, respectively, are necessary in these cases.

Geomorphological research is integrated into *Environmental Impact Assessment* (Gilpin 1995), a new discipline within environmental sciences with direct relevance to environmental management. In planning, various types of land use geomorphological information are of vital importance. Landscape evolution in the past, present and future is equally important, for instance, for waste disposal. To cite an example from Hungary, the site in the Geresd Hills envisaged for the disposal of low to medium activity radioactive waste of the Paks Nuclear Power Plant has been proved unsuitable for the purpose by geomorphologists (Balogh et al. 1990) since in this highly dissected hill area, the headward erosion of water-courses could consume narrow loess ridges before radiation reached a secure level. A better solution exists: the disposal site is now implemented in the Boda Siltstone Formation on the western margin of Mecsek Mountains, Southwest Hungary.

Another major contribution of process geomorphology to environmental management is observed in coastal areas. The human impact on coasts (e.g. development) endangers their sustainable use. Models are constructed to estimate how the changes in the sedimentation environment along the coast affect the pattern of erosion and accumulation. The risk of coastal erosion can be established and mitigation efforts can be planned. An example for calculations: along the Atlantic coast of the United States, the artificial sand recharge of beaches would cost ca USD 1,500,000 per mile!

3.6 Protecting Geomorphosites

Environmental geomorphology also covers the conservation of landforms. The motives for conservation can be numerous (Goldsmith 1983): ethical, spiritual, scientific, aesthetic, to maintain biodiversity, stability, sustainability, recreational, economic and preventing unintended impacts. The Working Group on 'Geomorphosites' of the International Association of Geomorphologists (IAG/AIG) makes efforts to have landforms recognized as part of the heritage of humanity. Nominations are also prepared for inclusion in the UNESCO List of World Natural Heritage. Landforms of anthropogenic origin (like terraced rice paddies and ancient burial mounds) are also candidates for heritage sites.

Not only landforms need protection (for instance, from the impact of increased tourism) but objects of cultural heritage may also be threatened by natural and man-induced geomorphic processes (soil erosion and accumulation, collapses of river bluffs with prehistoric earthworks and/or Roman castella, coastline retreat, etc.).

3.7 Land Reclamation and Restoration

The most intensively used landscapes of the Earth's surface usually become derelict after a time and need reclamation. The rehabilitation of the landscape is particularly important in *industrial* and *mining regions*. The approach has to be holistic, comprehensive of topography, drainage, microclimate, soils and vegetation. Spoil tips present special difficulties for restoration and stability has to be the first consideration. Disasters like the tragedy of Aberfan in 1966, when 115 schoolchildren became buried under 110,000 m^3 of rock (Bennett and Doyle 1997), warn of circumspection.

When designing topographic and drainage conditions slope inclinations and lengths must not reach the critical thresholds of soil erosion (Erdősi 1987). The values of these parameters reflect the equilibrium of accumulational and erosional processes. In order to minimize future soil erosion, the topographic conditions of the neighbouring (semi-natural) surfaces have to be analysed (Haigh 1978) and conclusions drawn.

In mining areas the drainage network has to be restored into semi-natural conditions (Lóczy et al. 2007). Pre-mining drainage density and future land use equally have to be considered. All components of water budget (precipitation, runoff ratio, evaporation, transpiration, deep infiltration, groundwater recharge) have to be taken into account. Climate change as well as the unpredictability of climate in some regions (like Central Europe), where it is extremely difficult to forecast, for instance, the duration and depth of annual snow cover, may modify these parameters in the near future.

Within the topic of land reclamation, the literature on the restoration of *riverine environment* is rapidly expanding (Brookes 1988). Rivers present problems to society but human interventions also cause serious problems to the functioning of river channel and floodplain ecosystems. Since the industrial revolution, the pressure on rivers has enormously increased (Table 3.2). A document of the US Department of Agriculture (Table 3.3 – USDA 2001) lists as many as 50 direct or indirect types of impact of 22 societal activities. In the United States, Germany and the United Kingdom, a number of previously channelized and straightened minor rivers were 're-meanderized'. River channel and floodplain restoration measures also serve landscape ecological purposes and create ecological corridors.

Recently, the idea arose that in certain cases river restoration is best implemented through the decommissioning of older dams of limited size (Lóczy 2001) and the operation of larger dams should be newly regulated. This way the fragmentation of fish habitats could be reduced. The contribution of geomorphologists to river restoration can be summarized as (Graf 1996):

– based on landscape evolution research, to establish what is natural;
– studying process to reveal the present-day operation of the system and
– to analyse the outcomes of various restoration options.

3.8 Perspectives

Obviously, in the 21st century geomorphology will acquire a more central role in environmental management. That the following three factors have to be in balance at decision making in environmental management also applies to geomorphological issues:

– *Politics* have to articulate demands from society.
– *Science* has to find the ways to satisfy these demands.
– *Economy* has to specify the resources available for the implementation of scientific solutions.

Finally, attention has to be paid to the requirements of sustainable development too.

3.9 Case Study 1: Water Transfer Schemes in China

Water resources are rather unevenly distributed in China. The North China Plain, the Shandong peninsula and the Loess Plateau contain 38% of China's arable land and only receive 6.7% of China's total water runoff. Using groundwater to meet demands for water for agriculture, industry and household consumption causes falling water tables and widespread subsidence under many cities. Rivers dry out in the rainless season. The idea of relieving this situation by transferring a large volume of water from the basin of the Changjiang (Yangtze) to the Huanghe (Yellow River) basin appears to have originated in the early 1950s and enjoyed Soviet support at that time but was abandoned for decades, only to revive in 1978 (Anonymous 1999). Then four alternative routes were considered. Since 2000, two of the canals are under construction and completion is foreseen to 2014. The third alternative, the 'Middle Route' is 1236 km long and partly draws water from the Danjiangkou reservoir built by 1974 and from the reservoir of the Sanxia (Three Gorges) Dam now ready. The fourth, now called the 'Eastern Route', which is designed to follow the old Grand Canal along 1130 km length and to conduct water from the lower Changjiang to Tianjin, is being simultaneously implemented. The expected environmental impacts of the canals are still debated. Interbasin water transfer along the Middle Route has to overcome considerable relief. The environmental impact of the Eastern Route may be deleterious because water may seep from the canal bed and raise the water table, thus increasing soil salinity and because it follows contaminated rivers.

3.10 Case Study 2: Geomorphological Information in Aggregate Exploration

In regions where Pleistocene ice sheets have produced typical assemblages of glacial and glacifluvial landforms, the individual landform types can be interpreted for sands and gravels exploration and geomorphological maps can be used for identifying reserves of these building materials. A British assessment technique (Crimes et al. 1992) simply lists the landforms of glacial/glacifluvial origin and attributes rank scores to them according to the quality of sand and gravel deposit to be gained from them. The technique was first tried successfully on the Llyn Peninsula, North Wales (Table 3.3). The typical composition of landforms is described from the review of geomorphological literature. The best potential sand and gravel reserves are contained in eskers, which are thus subjected to further investigation. The geomorphological map also clearly indicates outwash plains and kames formed towards the end of glaciation and they are perspective localities for sand and gravel extraction as well.

Table 3.3 Interpretation of the geomorphological map of a formerly glaciated area for sand and gravel deposits (modified and simplified from Crimes et al. 1992)

Landform	Characteristics	Assumed composition	Suitability
Till sheet (ground moraine plain)	Low relief, undulating plain with drumlins	Clay with disseminated clasts	*
Recession moraine	Accumulated at glacier snout during stillstand or readvance; arcuate or straight ridges, perpendicular to the former ice flow direction	Very variable: different types of till and glacifluvial sands/gravels	*
Eskers	Long, sinuous ridges accumulated from subglacial waters	Well sorted, cross-bedded aggregate, mixed with fines	*****
Kame hills	Mounds, hummocks, discontinuous terraces	Usually sand and gravel but also silt and clay, size limited	**
Kame terraces	Glacifluvial deposits between a glacier margin and the valley side	Usually dominated by sand and gravel, although silt and clay content may be high	***
Dead ice features	Complex assemblage of eskers, kames and irregular mounds deposited over buried ice	Variable composition, high silt and clay contents, investigation on the spot necessary	**
Sandur (outwash plain)	Extensive flat or fan-shaped proglacial accumulation of glacifluvial deposit	Large sand and gravel sheets, although variable in grain size and continuity	****

Suitability: * = poor to ***** = excellent

To the analogy of the above table, an exploration model can be suggested for countries where the main source of aggregates is fluvial deposits. Here too, the assessment is based on landforms identified on geomorphological maps (Table 3.4). In principle, the material dredged from still accumulating bars would be the most valuable but its amount is limited. Further investigations have to be performed to evaluate the impact of dredging on future channel evolution.

Table 3.4 Interpretation of geomorphological map for reserves of aggregates of fluvial deposition (compiled by Lóczy, based on the comparative analysis of Ryder and Howes 2001 and other sources)

Landforms	Characteristics	Assumed composition	Suitability
Point bar	Arcuate bars and swales in meander loops, below bankfull water level	Upward refining gravel and coarse sand, in swales silt	***
Mid-channel bar	In braided channel, thalweg on both sides	Well sorted, cross-bedded sand and gravel in larger extension	*****
Riffle	Shallow channel section in the inflexion belt of meandering river	Gravel and coarse sand	****
Pool	Local scours at uniform distance	Silt and fine sand in small extension	*
Natural levee	Floodplain deposition on both banks	Mainly medium to fine grained sand mixed with silt	**
Backswamp	Depression in distal floodplain	Horizontal clay and fine silt beds	*
Torrential delta	Braided channels at confluences	Well sorted, cross-bedded gravel and sand	***
Young terrace	Channel and floodplain deposition from penultimate and last glaciations	Cross-bedded gravel and sand, silt content <8%	****
Old terrace	More elevated channel and floodplain deposits older than penultimate glaciation	More weathered, cross-bedded gravel and sand, >8% fine grains	***

Suitability: * = poor to ***** = excellent

3.11 Case Study 3: Human-Induced Earthquakes

In the era of underground nuclear tests the explosion of a 1 Mt bomb resulted in a 6.9 M (Richter) earthquake and numerous aftershocks. In the Yucca Mountains, Nevada, 1 m dislocation was measured along a fault-line. Similar outcomes followed another kind of military operation. For 4 years beginning 1962, sewage of a US Army chemical plant in the Rocky Mountains was injected into a gneiss body at 4,000 m depth. Soon the more and more frequent quakes of 3–4 M size became a major concern in Denver, Colorado. When the connection between the events was disclosed, underground sewage disposal was stopped. The construction of dams and filling reservoirs behind them may also lead to earthquakes as it first became obvious in 1963, when tremors were observed in the vicinity of Lake Mead on the Colorado River. Dislocations may primarily result from pore pressure changes across fault surfaces which reduce shear strength and faulting reactivates (Goudie 2006). The relationship between river impounding and seismicity, however, is not so simple. There are indications that increased pore pressure may lead to intensified fault creep – thus reducing seismic activity (as observed in Canada – Milne and Berry 1976). The Tarbela Dam on the Indus in the North-West Frontier Province of Pakistan is another example of this trend.

References

Anonymous (1999) China South-North Water Transfer Project. Paper presented at East Asian Research Conference, University of Sheffield, July 1999 http://members.aol.com/anglochine/nsbd.htm

Balogh J, Schweitzer F, Tiner T (1990) Az Ófalu mellé tervezett radioaktívhulladék-temető földrajzi környezete (Geographical environment of the radioactive waste disposal site designed to be built near Ófalu). Földr Ért 39 (1–4): 103–131

Bennett MR, Doyle P (1997) Environmental Geology: Geology and the Human Environment. John Wiley and Sons, Chichester

Brookes A (1988) Channelized Rivers: Perspectives for Environmental Management. John Wiley and Sons, Chichester

Cooke RU, Doornkamp, JC (1990) Geomorphology in Environmental Management: A New Introduction. 2nd edn. Clarendon Press, Oxford

Crimes TP, Chester DK, Thomas GSP (1992) Exploration of sand and gravel resources by geomorphological analysis in the glacial sediments of the eastern Llyn Peninsula, Gwynedd, North Wales. Eng Geol 32: 137–156

Embleton C, Federici PR, Rodolfi G (1989) Geomorphological hazards. Supp Geogr Fís Dinam Quat II: 1–4

Erdősi F (1987) A társadalom hatása a felszínre, a vizekre és az éghajlatra a Mecsekben és tágabb térségében (Human Impact on the Surface, Waters and Climate in Mecsek MOUNTAINS and Environs). Akadémiai Kiadó, Budapest

Gerasimov IP (1968) Constructive geography: aims, methods and results. Sov Geogr Rev Trans 9:739–753

Gilbert GK (1917) Hydraulic Mining Debris in the Sierra Nevada. US Geological Survey Professional Paper 105

Gilpin A (1995) Environmental Impact Assessment: Cutting Edge for the Twenty-First Century. Cambridge University Press, Cambridge, UK

Goldsmith FB (1983) Evaluating nature. In: Warren A, Goldsmith FB (eds.), Conservation in Perspective. John Wiley and Sons, Chichester

Goudie AS (2006) The Human Impact on the Natural Environment. Blackwell, Oxford

Graf WL (1996) Geomorphology and policy for restoration of impounded American rivers: what is 'natural'? In: Rhoads BL, Thorn CE (eds.), The Scientific Nature of Geomorphology. John Wiley and Sons, New York

Gregory KJ (1985) The impact of river channelization. Geographical Journal 151: 53–74

Guzzetti F (2003) Landslide Cartography, Hazard Assessment and Risk Evaluation: Overview, Limits and Prospective. In: Mitigation of Climate Induced Natural Hazard Workshop 3. Proceedings, Wallingford, UK

Guzzetti F, Carrara A, Cardinali M, Reichenbach P (1999) Landslide hazard evaluation: a review of current techniques and their application in a multi-scale study. Geomorph 31: 181–216

Haigh MJ (1978) Evolution of Slopes on Artificial Landforms. Department of Geography, University of Chicago, Chicago. (Research Paper 183)

Hooke JM (ed.) (1988) Geomorphology in Environmental Planning. John Wiley and Sons, Chichester

Hudyma M (2004) Mining-Induced Seismicity in Underground, Mechanised, Hardrock Mines – Results of a World Wide Survey. Australian Centre for Geomechanics, The University of Western Australia, Nedlands, WA

Kelly PM, Campbell DA, Tarrant JR (1983) Large-scale water transfers in the USSR. Geo J 7 (3): 201–214

Lóczy D (2001) Folyóduzzasztás, víztározás – áldás vagy átok? (Damming rivers – a blessing or a curse?). In: Kovács J, Lóczy D (eds.), A vizek és az ember. Tiszteletkötet Lovász György Professzor Úr 70. születésnapjára (Waters and Man. Papers in Honour of Professor György Lovász on his 70th Birthday). PTE Institute of Geography, Pécs

Lóczy D, Czigány SZ, Dezső J, Gyenizse P, Kovács J, Nagyváradi L, Pirkhoffer E (2007) Geomorphological tasks in planning the rehabilitation of coal mining areas at Pécs, Hungary. Geogr Fís Dinam Quat 30: 203–207

Milne WG, Berry MJ (1976) Induced seismicity in Canada. Eng Geol 10: 219–226

Nagy L, Tóth S (2001) Veszély, zóna és kockázattérképezés (Mapping hazard, zones and risk).Vízügyi Közlemények 83 (2): 288–308

Panizza M (ed.) (1996) Environmental Geomorphology. Elsevier, Amsterdam. (Developments in Earth Surface Processes No. 4)

Robinson GD, Spieker AM (1978) Nature to be Commanded. – US Geological Survey Professional Paper 950

Ryder JM, Howes DE (2001) Terrain Information: A User's Guide to Terrain Maps in BC. In: Wildlife and Wildlife Habitat Inventory. Ministry of Sustainable Resource Management, Government of British Columbia, Victoria. http://srmwww.gov.bc.ca/rib/wis/terrain/publications/guide/index.html

USDA (2001) Restoration of Stream Corridors. US Department of Agriculture, Washington, DC. http://www.usda.gov/stream_restoration/PDFFILES. Pdf

Varnes DJ & IAEG (1984) Landslide hazard zonation – a review of principles and practice. – IAEG Commission on Landslides and other Mass Movements UNESCO, Paris

Chapter 4
Anthropogenic Geomorphology and Landscape Ecology

Péter Csorba

Abstract Since landscape ecology is the discipline of functionally studying natural factors and anthropogenic processes in light of the present and forecasted land-use tendencies, anthropogenic geomorphology easily fits in among the various fields of landscape ecology. The spatial distribution of human structures (built-up areas, roads, railways, channels and others) is always adjusted to topographic conditions. To rank the intensity of anthropogenic impact on a qualitative range, so-called hemeroby levels have been established by German scientists. When assessing hemeroby, estimations are made for the degree of human geomorphic impact based on the rate of soil erosion, surface dissection or the abundance of terraces, escarpments and artificial excavational features. At the highest level of human impact, in urban-industrial (or urban-technical) ecosystems, even remnant patches of semi-natural ecosystems seldom occur wedged into built-up areas and into linear infrastructural elements. The micro- and meso-elements of topography are often totally destroyed by terrain modification, such as levelling for development. Relying on anthropogenic geomorphology, landscape ecology can make significant practical contributions to landscape planning.

Keywords Landscape ecology · Hemeroby · Landscapes · Cultivated landscapes

4.1 Landscape Ecology as a Discipline

Landscape geographical research, since the 1960s, has increasingly acquired *an ecological approach* (Leser 1991; Finke 1986; Farina 1998; Csorba 2003; Wu and Hobbs 2007). In its simplest form, it means that phenomena and processes are studied *embedded in their environmental systems*. Recently the denomination

P. Csorba (✉)
Department of Landscape Protection and Environmental Geography, University of Debrecen, Egyetem tér 1, 4032 Debrecen, Hungary
e-mail: csorbap@delfin.unideb.hu

J. Szabó et al. (eds.), *Anthropogenic Geomorphology*,
DOI 10.1007/978-90-481-3058-0_4, © Springer Science+Business Media B.V. 2010

"*landscape research of ecological approach*" is used for this field of research. It is not much modified by the fact that the term "landscape ecology (or geoecology)" has become widespread in the international usage (Leser 1991; Huggett 1995). Among the fundamental characteristics of landscape ecology, a *practical approach* should also be accentuated (Helming and Wiggering 2003; Wiens and Moss 2005). Landscape ecology research primarily aims at fulfilling social demands in a way they should have the least pressure on potential natural resources and hinder the satisfaction of other social demands to the least possible extent. Landscape ecology provides a scientific background to achieve reasonable landscape management and land-use compromises (Marsh 1997; Ingegnoli 2002; Jongman 2005).

Landscape ecology, as a result of its roots in geography, also inherited the *spatial approach* of geography. A decisive question is where the various forms of social activities could be accommodated at the lowest physical-economic-social conflicts. According to Carl Troll, the founder of landscape ecology as an independent discipline (1939), landscape ecology is "Raumökologie der Erdoberfläche", i.e. the science of ecological processes on the Earth's surface.

Among the large number of definitions of landscape ecology, the following two are often cited:

> Landscape ecology is a science predestined to explore the diversity of spatial structures as well as to determine the place and possibilities of mankind (Neef 1984).
> Landscape ecology is the discipline of functionally studying natural factors and anthropogenic processes in light of the present and forecasted land-use tendencies (Naveh and Lieberman 1984).

Landscape ecology studies the landscape for the following purposes:

- structure,
- functioning and
- diurnal changes.

When the *subject, aim and methods* of landscape ecology are analysed in more details, it is seen that the *anthropogenic aspect* is central. Regarding its subject, e.g. it includes the research into agricultural or urban ecosystems, in terms of aims, e.g. it intends to increase the quality of human life and among its methods, the tools of social research, e.g. historical ecology, are also applied.

Anthropogenic geomorphology, consequently, easily fits among the various fields of landscape ecology, and the knowledge of ecological approached landscape research bulked during the last few decades would provide a useful theoretical background to anthropogenic geomorphology, too (Lóczy 2007).

4.2 Geomorphology and Landscape Ecology

Landscape ecology does not establish an order of importance among subsystems making up the landscape, i.e. topography, geologic-lithologic bedrock, climate, hydrology, soil, the living world and human activity, i.e. landscape is considered to be a *polycentric structure* (Haase 1999; Klopatek and Gardner 1999). This approach is fundamentally different from that of ecology, where research always focuses on the living being itself, or on its supra-individual organization levels (partial population, population, association, etc.). Therefore, ecology proper can be regarded as a discipline of monocentric approach (Csorba 2003).

The polycentric attitude of landscape ecology apparently does not mean that in the study of a given landscape, any of the above-listed factors would not be given pre-eminence (Mezősi and Rakonczai 1997). Only the dominant and subordinate elements are not always identical. There are landscapes with structure and function, where hydrology, while in others vegetation, has a predominant role. Taking a Hungarian example, for the functioning of the Bükk Mountains landscape, e.g. lithology (karstic limestone) is an essential factor, while in the Hortobágy, (alkaline) soils are relatively more important.

It seems certain, however, that *topography and human impact* can usually be found among the *drivers of landscape evolution*. Thus – in spite of the polycentric attitude of landscape research – the agents of anthropogenic geomorphology; i.e. topography and human activities, which shape it with extraordinary effectiveness, usually play a more important role in the structure and functioning of landscapes than the other landscape-forming factors (Grunert and Höllermann 1992).

Consequently, the topographic factor of the landscape and the geomorphic impact of social activities are generally integral elements of landscape ecological research, too.

Most of the landscape ecological research aim to determine landscape

- potential resources,
- stability,
- sensitivity,
- carrying capacity and
- diversity.

To answer the above questions, the role of topography in the structure and functioning of the landscape as well as in the history of its utilization has to be studied.

There has been a long debate in landscape ecological literature on what the landscape factors making a contribution to the *landscape persistence* (stability) are. According to what is acknowledged today by professionals, e.g. abiotic factors, by their heavy resistance, whereas biotic factors make a contribution to landscape stability through their flexibility. Topography is a factor less disposed to changes and belongs rather to the antecedent, more conservative landscape-forming factors. On the contrary, human impact is the most flexible landscape-forming force, the most quickly adjusting one to the external circumstances.

When the *role of topography* in the ecological *landscape structure* of the European cultural landscapes is examined, it can be concluded that, compared to topography, land use, geology and linear technical objects play a minor part in the shaping of landscape pattern. (Landscape pattern is the spatial arrangement of ecological patches, corridors and barriers – Forman 1995). Obviously, the spatial distribution of built-up areas, roads, railways, channels and other land use is adjusted to topographic conditions, although the ecological textures of landscapes are directly shaped by wood belts, plot boundaries, roadside fallow belts and openings for electricity transmission lines. This is coupled by the material and energy cycles of the landscape, habitat arrangement, ecological diversity of the landscape, in brief, by all aspects of landscape ecology (Forman 1995; Haines-Young 2005).

Another key question of landscape ecology is the definition of *landscape diversity* as well as the analysis of its temporal changes. Tendencies are mostly indicated by the changes in the mosaic-like character of land use. Authors claim that the land-use diversity of European landscapes peaked during the first half of the 19th century (Atkins et al. 1998; Wascher 2005). At that time, due to the increase in the number of inhabitants, all arable land was occupied by agriculture, most of the techniques correcting production sites through irrigation, fertilization, deep tillage were not yet sufficiently widespread to influence the structure and functioning of the landscape. In other words, land use was dominated by agriculture well adapted to habitat conditions. Between 1750 and 1850, at a number of locations in Europe some measures with geomorphological consequences, land reclamation, the stabilization of sand dunes, coasts and slopes, etc. began and had been present in dimensions never seen before until the second half of the 19th century. By the utilization of former floodplains, semi-fixed or wind-blown sand dunes, marshy, dune seashores for the purpose of silviculture, agriculture or grassland farming, there was a definite drop in the previous ecological or landscape ecological diversity. Among the most spectacular European examples of this process, the vast forest plantations in the Landes in southwestern France, reclamation of the countless peat bogs in the North German Plain, polders in the Netherlands or the arable lands reclaimed after river channelizations in the Great Hungarian Plain can be mentioned. In the 20th century, changes in land use almost everywhere reduced landscape diversity by the spreading of arable land and forest monocultures and plantations. Major and minor elements of the topography, as dry valleys, intermittent river beds, escarpments, alluvial fans, etc., have gradually disappeared within 100–150 years' time. Open-cast mining also had an impact on vast areas, especially in the eastern and southern parts of the continent, where urban expansion, the rapid sprawl of urban agglomerations, development of the suburban structure also contributed to this process.

The reduction of landscape diversity came to an end only in the 1990s and since then, by the increase in the rate of nature conservation and fallow areas, *the average European landscape is becoming more and more diverse*. This process is also related to the diversity of topography. The restoration of the meandering channels of minor watercourses channelized decades ago led to a significant increase in the diversity of several landscapes of Germany, the Netherlands and Switzerland. Also, the declaration of general protection for tumuli (so-called Cumanian mounds) in Hungary also promoted the conservation of the diversity of topography. The topographical component of the increase in landscape diversity is also impacted by the tendency urging the conservation of the elements of traditional land use all over Europe.

The best examples can be the cultural landscapes found on the list of UNESCO World Heritage sites. Most of the basic features of traditional land use intended to be revived and preserved are rooted exactly in the 18–19th century land use of great diversity that has been significantly altered by humans over the past 150 years. From Öland in Sweden to Tuscany in Italy, from Andalucia in Southern Spain to the Tokaj-Hegyalja Region in Hungary, the number of landscapes where complex landscape protection seems to be achieved not only in reservoir-like landscape sections but throughout the entire landscape, has been increasing. This is the result of a growth in landscape diversity, among others that of the diversity in topography, almost everywhere (Pedroli 2001; Wascher 2005).

4.3 Stages of Intensifying Human Impact on the Landscape

Due to the practical approach of landscape ecology, it intends to *confirm* its observations *by measurable data*. To rank the intensity of anthropogenic impact, so-called *hemeroby* levels have been established (Bastian and Schreiber 1994):

- ahemerobic = natural ecosystems,
- oligohemerobic = slightly modified ecosystems,
- mesohemerobic = semi-natural ecosystems,
- euhemerobic = ecosystems removed from nature,
- polyhemerobic = ecosystems alien to nature and
- metahemerobic = artificial ecosystems.

When referring to the above hemeroby levels, all landscape factors including topography, soil and land use are taken into account and the final rating represents their average. Among all landscape factors, the degree of human impact can be best measured for soils and vegetation. Thus, rating is taken place by, for instance, changes in soil pH, the degree of alteration in the composition of elements due to fertilization and the use of chemicals, or in the case of vegetation cover, by the percentage of neophytic species (from the Americas or from Australia). Unfortunately, no such relatively well-applicable indicator is available for topography, at the best only estimations can be made for the degree of human impact based on the rate of erosion of the soil cover, surface dissection or the abundance of terraces, escarpments and artificial excavations.

4.3.1 Natural Landscapes

Their functioning is not directly influenced by human impact, thus the landscape is basically a self-regulatory system. Such systems are called natural (bio) ecosystems in ecology.

In Europe, only the northernmost, subarctic and high mountainous landscape sections of small extension can be classified into this category. Most of the national parks are excluded.

Anthropogenic impact, here, is enforced by air and water pollution (seas and rivers), impacting topography only in an indirect way (e.g. by the impact of acidic deposition influencing weathering and modifying debris generation). Locally, mining activity, transhumance animal husbandry, recently tourism has increasingly become the main factor of environmental disruption. In addition, the exploitation of energy sources, ores, mineral resources and construction materials has a direct modifying impact on topography, such as on the Kola Peninsula (phosphates), around Kiruna (iron ore) or Vorkuta (coal). With increasing environmental awareness, animal husbandry in this zone (mainly reindeer husbandry) seems to decline as a pressure on the natural system, while tourism, which is becoming a fundamental social demand, probably represents the most serious threat to the highly susceptible subarctic and mountain landscapes. Treading, rock climbing, skiing, mountain biking result in significant degradation of topography even at parts of the European mountain regions above timber line that are difficult to access, and in the subarctic zone (Frislid 1990). This landscape type disappeared from Central Europe already centuries ago.

4.3.2 Slightly Modified Landscapes

In such landscapes, there is only a minor human impact, after which the landscape system is capable of almost perfect recovery within a short period of time and regains its ability for self-regulation.

This type includes mostly sparsely populated (rural) North-European, the most arid southern and southeastern peninsulas and islands, the technologically influenced ecosystems of mountain regions (mechanized pastureland management and silviculture) as well as agricultural and silvicultural regions, where the principles of sustainable ecological farming are observed (Wascher and Jongman 2000).

Environmental pressure is randomly distributed in time as it can be more intense as a consequence of national or European Union rural development project, however, the consecutive dereliction in such areas is typical. Topography is often transformed to the highest degree by large-scale hydro-power projects, such as in Scandinavia or in the Alps (Plates 4.1 and 4.2).

To halt the depopulation of areas unsuitable for intensive agriculture, major efforts are made by the European economic policy. Sustainable landscape management is targeted by a liberal support system on the one hand and by manufacturing products fitting best to the ecological conditions of the landscape, e.g. collecting herbs, animal husbandry, on the other hand. The intensity of human impact, however, is not reduced by the fact that a significant population retaining influence is intended to be devoted to rural tourism, assuring nearly half of the necessary incomes from such alternative activities. This is especially true for mountainous

Plate 4.1 Severe transformation of the topography in an otherwise only slightly modified natural environment. Car parks in the Dolomites (Tre Cime di Lavaredo, Italy) (Csorba 2005)

Plate 4.2 Landscape with an apparently quasi-natural topography where, practically, the surface was also transformed to a large scale during the construction of a golf course. (Tale, Lower Tatra, Slovakia) (Csorba 2004)

areas along some coasts of Southern Europe, on the Greek Islands and Sicily. Environmental pressure is mosaical here. For instance, there are factory livestock sites, intensive silvicultural properties and some overcrowded tourism destinations where even the original topography has undergone significant changes while most of the area is derelict. From the point of view of landscape ecology, this sharp contrast is a characteristic feature of the landscapes mentioned above (Pedroli et al. 2007).

A realistic objective to be achieved is, in general, as regarded by European national parks is this quasi-natural ecological stage. In Hungary's national parks, such quasi-natural landscape type can be found in the Kiskunság, Hortobágy and Aggtelek National Parks. (Locally increased anthropogenic pressure, however, can be detected here as well, around visitors' centres and nature trails, where disturbance far exceeds the level of the strictly protected biosphere reserves in the core areas.)

4.3.3 Semi-natural Landscapes

With the decline of human use, the original physical conditions of such landscapes can be restored as topography (e.g. by constructing terraces), soil (e.g. by secondary alkalization), water balance (e.g. by water regulation) and microclimate (e.g. by development) have undergone enduring changes. Landscapes included in this category – converted and modified to a considerable degree – are mentioned by the German literature on landscape ecology as "manipulated" landscapes (Bastian and Schreiber 1994). The ability of the landscape for self-regulation can only be renewed in its modified form, restoration of the former conditions can only be achieved exclusively by conscious ecological landscape planning over a longer period of time (Head 2000; Mitchell and Ryan 2001).

Such landscapes are called semi-natural as in their functioning and appearance, ecosystems resembling to natural ones still predominate. The ecological functions of forest plantations with the expansion of the foliage and undergrowth, the provision of habitat for birds and insects, etc. are still rather close to the conditions prevailing in natural forests; pastures treated with herbicides can be regarded grasslands, arable fields also provide coverage at surfaces previously dominated by photosynthetizing green phytomass. The share of built-up areas in such landscapes does not exceed 20–25%, however the network of infrastructure is rather dense (road and railway cuts, channels, electricity transmission lines, shelter belts, etc.) having a severe so-called fragmentation impact on habitats (Forman 1995).

Most of Europe's area is included in this landscape category. Practically, it is yet loosely built-up cultivated landscapes, which replaced deciduous forests and grassland ecosystems (Atkins et al. 1998; Richling et al. 1998; Pedroli et al. 2007).

If the categories valuable for nature conservation are considered, this type could be most appropriately named "protected landscapes". The impacts of human land use is significant and well visible everywhere, although the rate of alien artificial surfaces is low as well as serious interventions to landscape functions (e.g. construction of water reservoirs, motorways, intensive farming around ecologically valuable habitat relicts, etc.) are prohibited. More and more areas have been declared protected landscapes (so-called nature parks) in Western Europe (Mander and Jongman 2000).

One of the main fields of European landscape protection is cultivated landscapes. It aims to stabilize the functioning of such landscapes, the resultant spatial structure and pattern. The role of cultivated landscapes is relevant as national identity is linked to them, thus a country, a region or a group of settlements can object or consciously oppose against their loss. Here, actually *landscape protection* is conceived in its broadest sense. Communities often support, even financially, the preservation of the alignment of roads, the species composition of roadside alleys, the scenery of settlements, terrace systems, irrigation canals and traditional agricultural land use patterns. Such complex landscape protection programs are realized in Andalucia in Spain, in Languedoc-Roussillon in Southern France and in Tuscany in Italy. On the map of Europe's landscape types, mainly these regional structures have a privileged position, e.g. the so-called bocage (semi-opened grassland landscape jointed by smaller shrubby forest patches, where plot boundaries are marked by green hedges or stone walls) in Bretagne, kampen (another semi-opened type, with a higher share of arable land) along the border between Belgium and the Netherlands, the so-called coltura promiscua (agricultural area with mixed farming with the dominance of vine, olive and grains) in Central and Northern Italy and the so-called silvo-pastoral type (olive-wood plantation where grass remaining under the protection of the foliage is grazed) in Castile in Spain (Wascher 2000, 2005). The preservation of these as well as the expansion of their areas is encouraged by considerable financial support from the European Union (Pedroli et al. 2007).

Human impact on the topography in this category is of intermediate scale. The most evident landforms of human origin are terrace systems established for plantations, reconstructions of the beds of minor watercourses, drainage of the waters of swampy depressions, the abolishment of minor marshes, bogs and ponds. Roads and, especially, railways are directed either on banks or in cuts, therefore, slopes, viaducts and tunnels locally exhibit a significant impact on the landscape. Such new elements of the topography may have a considerable modifying impact on surface runoff, even a remarkable influence on subsurface runoff, as well as on microclimate. Most of the surface still preserves its original character, predominant features of the scenery are usually mosaical agricultural fields adjusted to the topography. (Surfaces with more than two-thirds of them occupied by large continuous terraced slopes are excluded from this category. Such are, for instance, the Kaiserstuhl in the Rhine Valley, Cinque Terre on the Ligurian coast or the surroundings of Tokaj in Hungary. There is a greater anthropogenic transformation in the case of polders in the Netherlands as there every square meter of land is created by humans.)

In Hungary, too, the predominance of such semi-natural landscape types – where the more stable landscape factors, including topography, still bear most of its former natural features – is experienced (Csorba 2002). Practically, it includes most of the agricultural lowland and hill landscapes with a sparser network of settlements. This is obviously larger than the official protected landscapes and, if the category "nature parks" were introduced, it would cover 50–60% of the country's territory.

4.3.4 Landscapes Removed From and Alien to Natural Conditions

The functioning of landscapes classified into this category is driven by permanent human activity almost everywhere (Pedroli 2001). The landscape is not yet entirely built-up, with a densely populated, though not yet agglomerating, settlement network. The areas of the most intensively cultivated arable lands, gardens and vast plantations belong to this category, from the wine monoculture of the Duero Valley, through the tulip fields in the Netherlands to the rice lands in Lombardy, to the "greenhouse landscapes" in southern Spain as well as the stable Alpine animal husbandry, which artificially improves ecological conditions. (Dairy, vegetables, fruit and other horticultural supply belt formed around the largest agglomerations is not included here as they lie in suburban belts where the pattern and functioning of the landscape is dependent on the ratio of areas developed for housing and infrastructure.)

Instead of the usual originally mosaical arrangement of natural features, a homogenizing impact of humans is noticeable (Plate 4.3). At many locations, practically monocultures can be found along with all their ecological drawbacks. Water supply is artificial (irrigated cultivation), the modification of topography is substantial (terraces, channels), landscape ecological relationships have been restricted to a minimum (incomplete network of green corridors). Semi-natural habitat patches are

Plate 4.3 Well-managed cultural landscape; the upper valley of Salzach (St. Georgen, Austria). Fine details of the original topography have been obliterated by agriculture; the river has been redirected to a channelized bed. (Csorba 2005)

usually isolated, stepping stone type relicts. Nature conservation strives for conservation but only consolidation and stabilization of ecotopes slowing down the process of final decay can be expected.

This landscape type is typical in certain parts of France (in the Loire Valley), large areas in Germany (Southern Bavaria), most of the Benelux States (Flanders, Brabant), landscapes of the former Black Country in England as well as in the loosely built-up coastal and mountain resort districts (central Provence, the densely populated valleys of Tirol, the North Italian lake district, etc.).

Among the most typical agricultural landscapes, huertas in the Mediterranean, the polders in the Netherlands, the rosaries of the Tunja Valley in Bulgaria or some renowned tobacco plantations in Northern Greece exhibit very specialized landscape management. In Hungary, some wine-growing regions could only be classified to this type. Due to the fallback of production over the past decades, the higher rate of forests and the not-remarkably dense settlement network in the apple-growing area of the Nyírség Region does not fit here, whereas the country's famous paprika and onion-growing areas of limited extension do not constitute landscapes proper.

Positive examples for the successful development of semi-natural habitats or habitats removed from natural can be found where the sustenance of the last isolates serve the interests of the European ecological network or where cross-border ecological connections depend on them. Such preserved ecological green corridor is the lower section of the River Rhine along the border between Germany and the Netherlands (Herget and Dikau 2005). The increasing role of environmental protection is represented by the growing rate of protected areas along border rivers or mountains, and in the past 10–15 years, gestures are made to demonstrate willingness for the sake of the common interests of the unifying Europe.

4.3.5 Anthropogenic Landscapes

In these urban-industrial (or urban-technical) ecosystems even remnant patches of ecosystems removed from natural seldom occur, only semi-natural ecosystem fragments are wedged into the spatial structure densely jointed by built-up areas and linear infrastructural elements (Plate 4.4) (Mühlenberg and Slowik 1997).

Due to the high realty prices of the agglomeration zone of cities, only agricultural production of the highest value can keep its position against the growth of the city, by which it is furthered. During the past decades, elementary social demand on recreation and proximity to nature seem to compete with agriculture. Modern urban planning intends to extend the rate of green areas, and even, to develop urban and yet undeveloped suburban areas into a possibly liveable network of interconnected elements.

From the point of view of ecology, such hybrid urban-industrial-green area ecosystems are rather reduced living systems sustained and operated by horticulture. First, leaf litter is removed and then the resulting suppressed organic substance is artificially replaced. Most of the surface is sealed, rainwater runs into channels

Plate 4.4 Due to the urban built-up, only the sharpest features of the original topography are predominant. (The City of Lisboa from the St. George's Castle, Portugal) (Csorba 2005)

then onto green areas and trees are irrigated for months. Species composition is much more determined by tolerance to disturbance and contaminants than the features of the original flora. Fauna has an absolutely reduced abnormal composition; food chain is fairly simple and incomplete.

The micro- and meso-elements of topography have been totally destroyed by terrain modification, such as levelling for development. The topography is regarded an obstacle to construction and regulation and stabilization are aimed to be achieved by serious interventions of engineering geology. The only positive feature of topography has remained to provide fine outlook to the environment from downhill terrains – also reflected in realty prices.

Most of Europe's large agglomerations developed during the Industrial Revolution (the Black County in England, the Ruhr District, Belgium's industrial area, etc.) in the past 40–50 years have greatly overpassed a development phase that associated their names with unhealthy life conditions. Today, in the towns of the Ruhr District, for instance, hardly any smoking high chimneys can be found, and the agglomerating towns are isolated by greenbelts. In addition to the classic industrial agglomerations mentioned, mostly European cities with a population exceeding 2 million also have a landscape forming role. Suburban ecological spatial structure from Madrid to Munich, from Moscow to Brussels are sharply formed by ring roads, while airports, railway freight traffic, warehouses, water treatment plants and waste deposition sites all have a great land demand. Among sharp land-use conflicts, nature conservation interests can only predominate when natural values are successfully represented by market value categories.

There are rare examples for former suburban warehouses, now empty, being transformed into green areas for recreation. Even though, we are aware that the ecological value of such park landscapes is still far behind those of modified natural ecosystems, unfortunately such cases are still exceptions.

4.4 Summary

While under European conditions, a more and more comfortable settlement with rich infrastructure is typical, demand for an almost natural, harmonic landscape experience is also widely present. An increasing significance is attributed to a secure, pollution-free, quiet environment for everyday life. To this purpose landscapes have been substantially modified. Such interventions, even on the level of landscapes, may have a serious impact on one of the most stable elements of landscapes, i.e. topography. Relying on anthropogenic geomorphology and through its system-based approach, landscape ecology can make a significant contribution to practice.

References

Atkins P, Simmons I, Roberts B (1998) People, Land and Time. An Historical Introduction to the Relations Between Landscapes Culture and Environment. Arnold, London

Bastian O, Schreiber K-F (1994) Analyse und ökologische Bewertung der Landschaft. G. Fischer VerlagGerman

Csorba P (2002) Agricultural Land Use History of the Great Hungarian Plain, with Special Attention to the Hortobágy. Oldenburger Geoökologische Studien Band 5

Csorba P (2003) Tájökológia. Kossuth Egyetemi Kiadó, Debrecen

Farina A (1998) Principles and Methods in Landscape Ecology. Chapman and Hall, London

Finke L (1986) Landschaftsökologie. Höller-Zwick, Westermann

Forman RTT (1995) Land Mosaics. The Ecology of Landscapes and Regions. Cambridge University Press, Cambridge

Frislid R (1990) Cultural Landscapes of Norway. Landbruksforlaget

Grunert J, Höllermann P (eds.) (1992) Geomorphologie und Landschaftsökologie. Bonner Geographische Abhandlungen, Heft 85.

Haase G (1999) Beiträge zur Landschaftsanalyse und Landschaftsdiagnose. Abhandlungen der Sächsische Akademie der Wissenschaft zu Leipzig, 59, Heft 1

Haines-Young R (2005) Landscape pattern context and process. In: Wiens J, Moss, M (eds.), Issues and Perspectives in Landscape Ecology. Cambridge Studies in Landscape Ecology, Cambridge University PressCambridge

Head L (2000) Cultural Landscapes and Environmental Change. Arnold, London

Helming K, Wiggering H (eds.) (2003) Sustainable Development of Multifunctional Landscapes. Springer Verlag, Berlin, Heidelberg, New York

Herget J, Dikau R (eds.) (2005) Natural and human impacts in the River Rhine Catchment. Erdkunde 59, Heft 3/4 (special issue)

Huggett J (1995) Geoecology. Routledge, London

Ingegnoli V (2002) Landscape Ecology: A Widening Foundation. Springer, Verlag

Jongman R (2005) Landscape ecology in land-use planning. In: Wiens, J, Moss, M (eds.), 2005: Issues and Perspectives in Landscape Ecology. Cambridge Studies in Landscape Ecology, Cambridge University Press, Cambridge

Klopatek JM, Gardner RH (eds.) (1999) Landscape Ecological Analysis. Springer, Verlag
Leser H (1991) Landschaftsökologie. UTB 521, Ulmer Verlag, Stuttgart
Lóczy D (2007) Predictable trends for geomorphological and landscape research. (in Hungarian) Földrajzi Közlemények 132: 1
Mander Ü, Jongman R (eds.) (2000) Consequences of Land Use Changes. WIT Press, Europe
Marsh W. M (1997) Landscape Planning: Environmental Application. John Wiley and Sons, New York
Mezősi G, Rakonczai J (szerk.) (1997) A geoökológiai térképezés elmélete és gyakorlata. JATE, Szeged
Mitchell F, Ryan M (2001) Reading the Irish Landscape. Town House, Dublin
Mühlenberg M, Slowik J (1997) Kulturlandschaft als Lebensraum. UTB 1947, Quelle & Meyer Verlag, Wiesbaden
Naveh Z, Lieberman AS (1984) Landscape Ecology. Theory and Application. Springer, Verlag
Neef E (1984) Emil Neef Planned Presentation for the First Congress of IALE. Roskilde University, Geography
Pedroli B (ed.) (2001) Landscape- Our Home Lebensraum Landschaft. Essays on the Culture on the European Landscape as a Task. Indigo Freies Geistesleben, Zeist
Pedroli B, Van Doorn A, De Blust G, Paracchini ML, Wascher D, Bunce F (eds.) (2007) Europe's Living Landscapes. Essays Exporing Our Identity in the Countryside. Landscape Europe, Wageningen/KNNV Publishing, Zeist
Richling A, Lechnio J, Malinowska E (eds.) (1998) Landscape Transformation in Europe. Practical and Theorethical Aspects. IALE Polish Association, Warsaw University
Wascher D (ed.) (2000) The Face of Europe. Policy Perspectives for European Landscapes. ECNC, Technical report series
Wascher D (ed.) (2005) European landscape Character Areas. Typologies, Cartography and Indicators for the Assessment of Sustainable Landscapes. Landscape Europe, Alterra Repost, No. 1254
Wascher D, Jongman R (2000) European Landscapes. Classification, Assessment and Conservation. European Environmental Agency, Coppenhagen
Wiens J, Moss M (eds.) (2005) Issues and Perspectives in Landscape Ecology. Cambridge Studies in Landscape Ecology. Cambridge University Press, Cambridge
Wu J, Hobbs RJ (eds.) (2007) Key Topics in Landscape Ecology. Cambridge Studies in Landscape Ecology. Cambridge University Press, Cambridge

Part III
Impacts of Various Human Activities on the Landscape

Chapter 5
Agriculture: Crop Cultivation and Horticulture

József Lóki

Abstract Alteration to the surface began with hoed cultivation; it is thus claimed that crop cultivation is the oldest production activity with geomorphic impact. The physical and social factors that influence the type, features, density, etc. of landforms resultant from agricultural activities are summarized. Crop cultivation has its main impact on the planation of the surface. Sloping surfaces impede cultivation, thus humans often reduced slope gradient or established horizontal plots. The relief of gently sloping areas gradually decreases by contour tillage. In areas with no soil conservation (such as in the Mediterranean) bare rocks indicate the full degradation of soils. In large-scale farming of mechanized cultivation, surface levelling is common. Mainly large-scale farmlands are also affected by wind erosion. In arid regions with water deficit, irrigated crop cultivation has been practised for ages and irrigation canals have formed an integral part of the agricultural landscape. Tillage modifies soil erodibility (slope, soil moisture, soil water management and structure, surface roughness and coverage) and thus water erosion, particularly on arable lands with soils on loess parent material.

Keywords Cultivation · Irrigation · Tillage · Water erosion · Wind-blown sand areas · Soil erosion

5.1 A Historical Review of Crop Cultivation

Humans have ever lived in communities of various sizes. In the early period of human history, settlements with favourable physical conditions to satisfy their demands were chosen. Vital food was gained by *collecting, fishing and hunting.*

J. Lóki (✉)
Department of Physical Geography and Geoinformatics, University of Debrecen, Egyetem tér 1, 4032 Debrecen, Hungary
e-mail: jloki@delfin.klte.hu

J. Szabó et al. (eds.), *Anthropogenic Geomorphology*,
DOI 10.1007/978-90-481-3058-0_5, © Springer Science+Business Media B.V. 2010

According to hypotheses (Láng 1961), the number of plant types known and consumed at that time was ca 2,700. This collecting, fishing and hunting way of living made humankind entirely dependent on nature and provided barely adequate means of nutrition. Residents remained in the area chosen until the physical environment could fulfil their demand and until they felt secure. When the conditions changed, they sought for a new site of settlement. This did not seem to be a problem as the Earth was only sparsely populated.

Collecting food was later replaced by cultivation of the land with the hoe and then with the plough. *Hoe farming* dates back to a period several thousand years ago. Characteristic tools of the hoe culture are the digging stick, the foot-plough, the spade and the axe. Fields were not manured; therefore soils with reduced productivity were left fallow and further areas were cleared of trees and bushes either by using axes or embroiling. Even today, there are regions where such primitive farming is still practised (Plate 5.1).

Alteration to the surface had basically begun with hoe cultivation; it can thus be claimed that crop cultivation is the oldest production activity with geomorphic impact.

Ploughing meant a great progress compared to hoe cultivation, as areas previously unutilized could also be involved in crop cultivation. Consequently, crop cultivation extended to the temperate zone, where human impact on the surface is of a large scale ever since. Primitive ploughs (Plate 5.2) permitted topsoil turning only; the degradation of the surface, however, increased over larger areas.

In the past decades, the surface has undergone significant alteration in the areas cultivated. A number of microforms (e.g. cuts, drainage channels), which change from year to year as a result of human intervention, can be mentioned along with

Plate 5.1 Hoe land cultivation in Africa (source: http://cozay.com/documents/)

Plate 5.2 Land cultivated by a wooden plough (source: http://exploringafrica.matrix.msu.edu/students/images/)

man-made and later abandoned landforms, altered or obliterated by ensuing physical processes (Erdősi 1969, 1987).

A review on landforms still observable today, regardless the time of their formation (even today), is provided below. Some of the landforms are the direct results

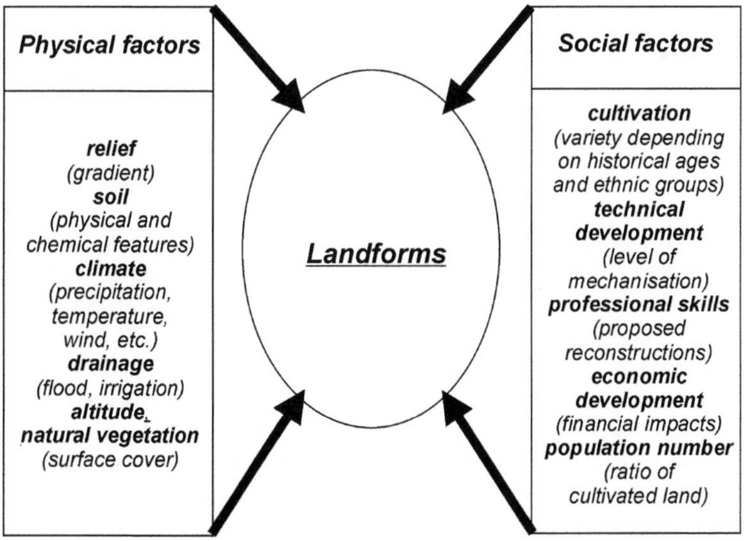

Fig. 5.1 Factors influencing landforms resultant from agriculture and horticulture (by Lóki in Szabó and Dávid 2006)

of crop cultivation or horticulture (e.g. terraces), while others are indirect consequences of farming (e.g. landforms of wind erosion). On arable land, natural geomorphic processes are often intensified or impeded by tillage.

Physical and social factors both influence the type, features, number, etc. of landforms resultant from agricultural activities (Fig. 5.1).

5.2 The Impact of Crop Cultivation and Horticulture on the Surface

Crop cultivation takes place primarily in plains and areas with gentle slopes, whereas steeper slopes can also be the sites of *horticulture* (viticulture and pomiculture). *Slope conditions* influence, on the one hand, the evolution of landforms and, on the other, the natural transformation of landforms of human origin. In the cultivation of lowlands, the course of positive (ridges) and negative (plough furrows) landforms formed by a plough is perpendicular to the route leading to the plot. When dividing plots and designing *farm-road pattern*, social factors (interests) were primarily taken into consideration. In the case of horticulture expanded to hillsides, the impact of social factors is apparent in the shaping of plots, although for soil conservation contour ploughing or terracing is a priority in many locations. Consequently, tillage techniques are reflected in the alignment of various landforms.

5.2.1 Influencing Factors

Soil fertility depends on physical and chemical soil properties. In addition to the expansion of crop cultivation, soil type controls the dimensions and lifetime of *landforms*. Crop production first gained ground on floodplain soils providing high yields and being easy to cultivate. With increasing demand for food, advantage was taken of technological development (social impact), and cultivation extended over areas harder to cultivate and of lower quality; consequently, human-induced landforms became apparent. Microforms developed by ploughing are smaller in size and tend to disappear sooner on loose sandy soils than on highly cohesive soils with high loam and clay content.

Climatic factors are mostly significant in the formation of primary anthropogenic landforms. The heat and rainwater demands of crops determine the spatial expansion of crop production, thus these also influence the distribution of crops cultivated. The various types of cultivation by plot size, the technology applied, etc. are also reflected in primary landforms and can also promote the emergence of secondary landforms (e.g. large-sized plots – deflation landforms).

Excess water (floods, inland water) and water deficit (compensated by irrigation) both have an impact on landforms of agricultural origin. Crop production began on fertile floodplains, but year by year floods also resulted in the disappearance of agricultural landforms. Crop cultivation can only be practised in wet lowlands if they are drained. In areas with water shortage, the establishment of irrigation canals allows cultivation.

During crop cultivation and horticulture, the types of plants grown are determined by climatic conditions (temperature, precipitation, etc.). As the conditions for production also change with *elevation*, the occurrence of agricultural landforms is limited by altitude. This limitation varies according to climatic zones. In lower geographical latitudes, landforms related to crop cultivation and horticulture also occur in the higher zones of mountains. Cultivation is also determined by *slope gradient* and *exposure*. The use of sloping land also depends on the technique applied in cultivation. Hoe horticulture can be used for steep slopes, whereas mechanical cultivation can be used only for gentle slopes.

Primary (forest or grassland) *vegetation* provides protection for the surface against degradation. Deforestation, the conversion of grasslands to crop cultivation and horticulture causes an increase in the intensity of soil degradation.

In addition to the climate, hydrology and relief, *soil quality* and fertility also contributed to the expansion of humanity, social progress and human alteration of the surface. The spatial expansion of crop cultivation is controlled by *social demand* as well as physical factors. In old times, soil suitability for food production was regarded to be an important factor which decided if particular soil types are cultivated or not. The rapid growth of population necessitated the cultivation of ever more extensive areas. To gain more ground, tropical rainforests were cleared, desert or semi-desert areas were involved in cultivation. Today, approximately 11% of terrestrial areas are cultivated (arable land, gardens, vineyards, orchards). An increase of arable land can be forecast where, on the one hand, there is a large population and, on the other, where either the conditions are provided or can be facilitated by applying a more advanced *technology*. Technological development, from the wooden plough to the high-capacity power machines, always had an influence not only on the increase of crop productivity, but also on geomorphic evolution.

5.2.2 Human Interventions and Human-induced Processes and Features

Landforms deriving from crop cultivation – as discussed in Chapter 1 – can be primary, secondary and can be formed either in a direct or indirect way.

Crop cultivation has its main impact on the *planation* of the surface. Slope surfaces impede both manual and mechanical works, thus humans often directly intend to decrease the slope gradient or establish horizontal plots. The relief of gently sloping areas gradually decreases by contour tillage. Where the slope is divided by plots of various cultivation, annually *growing steps* along plot boundaries (Ackerterrassen in German) can develop. The development of similar landforms can be observed along pathways and farm roads running downslope. Such landforms are often mentioned as pseudo-terraces (Szabó 1993). Alleys, shrubs and even fences along plot boundaries increase relief by retaining the downward wash of soil. On cultivated land, excavation is typical upslope and accumulation downslope, while basically a slow levelling process takes place. This process, at many locations, is also accelerated by *terracing*. Terraced cultivation has been applied from the earliest times on several continents (as discussed in detail in Chapter 7).

Ridges developed by tillage can be obliterated by both human intervention (e.g. dragging) and physical processes (physical or chemical weathering, sheet wash erosion). Therefore, runoff from rainwater or snow-melt causes *gully erosion* and secondary landforms (Plate 5.3). They appear on slopes as excavation landforms; in the footslope zone, however, the transported debris accumulates. With no soil conservation, surface degradation and slope alteration ensues. Such processes are apparent in various regions of the world: in the steppes of Ukraine and southern Russia the resulting features are called ovrags, in North America gully systems or arroyos. If topsoil removal and surface degradation extend over larger areas, badlands develop.

In mountains with a shallow soil cover, cultivated soils are easily eroded and *bedrock* becomes *exposed*. First, smaller or larger weathered pieces of the bedrock appear in the thinning soil layer (Plate 5.4) and impede tillage. The boulders are collected along the plot boundaries.

Rock bunds built crosswise on the slope can abate the complete degradation of soils. They can be regarded as primary positive landforms. Rock bunds in the slope direction are classified to be secondary landforms, as in their case the main aim was not to build this landform but to reduce the obstacles to cultivation. In areas with no soil conservation or where the prevention carried out proved to be inadequate, bare rocks indicate the full degradation of soils, as in the barren karst regions of the Mediterranean cultivated for millennia (e.g. Greece, Croatia, Dalmatia, Istria).

Plate 5.3 Rainwater erosion on a tilled slope (source: http://www.kwaad.net/Photo F. Kwaad)

Plate 5.4 Rock bunds along the plot boundary (source: taken by Szabó (2001))

The case is similar to other karst areas where scars cropping out or being subsurface can be chipped off by tillage (Fig. 5.2).

Rock bunds and piles can be seen in the area of the German–Polish Plains as well. In the moraine accumulated during the Pleistocene, boulders often several tens of centimetres in diameter commonly impede crop cultivation and they are exposed by ploughing year by year. These secondary accumulation landforms are made up by rocks collected over a long period of time. Where the intension was to form larger plots, the rocks collected were either carried away or piled up along the plot boundary and farm roads.

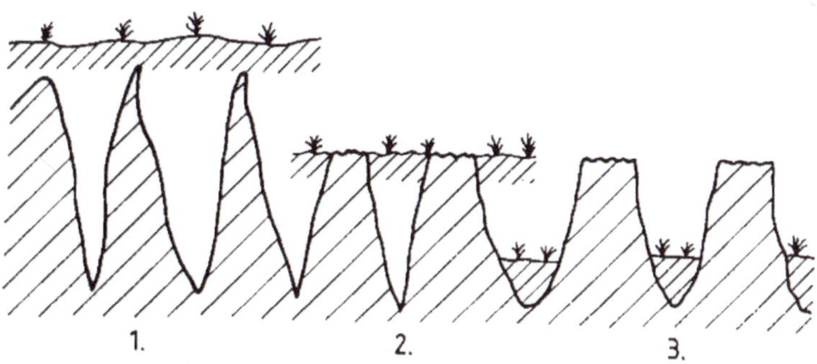

Fig. 5.2 Alterations in the surface of karstic areas by tillage 1 = prior to the beginning of cultivation, 2 = truncated scars at the time of cultivation, 3 = soil surface after cultivation stopped (according to Rathjens 1979)

In large-scale farming, surface levelling by mechanical cultivation is common. It was practised in the last century mostly in areas of wind-blown sand in Hungary (Danube–Tisza Interfluve, Nyírség, Inner Somogy) during the establishment of large plots of state farms and cooperatives following World War II. Levelling aimed at backfilling the low inter-hillock surfaces by sand from the top of these *hillocks*.

On the surfaces of wind-blown sand regions, no soil cover of high quality could develop and the hillock layer has been destroyed. In the Nyírség Region where hillocks are usually jointed by clay illuviation layers to a depth of several metres, serious damage has been caused. These layers, mainly due to their fine granulometric composition have a good moisture storage capacity, which makes them suitable for both crop production and fruit growing. By having the layers *stripped*, the water budget changed. The reduction in soil water storage capacity promotes wind erosion.

When *establishing large plots* of several tens or hundreds of hectares, the previous heterogeneous small- and medium-sized plot systems, well adapted to site conditions, were homogenised. Forest and shrub *shelter belts* were *removed*; their turf, ditches and ox-bow lakes were levelled by tillage. As a result, the natural mosaical nature of the landscape has disappeared and heterogeneous soil conditions unified. These alleys, ploughed turfs and wetlands played an extremely important role in the control of microclimate on arable land. With their disappearance, the micro- and mesoclimate have become drier and more extreme, soil water-retaining capacity has been reduced, and subsequent damage caused by droughts has become more frequent.

5.2.3 Wind Erosion and Its Impacts

Large-scale farmlands are affected by *wind erosion* mainly in early spring, but it is also apparent when autumns and winters are dry. By the re-deposition of the surface layer, the removal of humus colloids, soil fertility decreases. The dust transported as well as fertiliser and chemical residues blown out along with soil particles also mean a threat for human health. Wind erosion impacts are also apparent in areas with more cohesive soils. Intensive mechanical tillage by power machines removes the natural layer of soils and *destroys soil aggregates*. As a result, the top, ploughed layer of the soil becomes dusty, its water budget and ventilation deteriorate. Fine soil dust is removed and transported from the barren soil surface by the wind that finds no obstacles in the open landscape. This soil, easily desiccating and of permanently barren surface, can be easily blown away by the wind. Stronger winds can transport 10–100 tonnes per hectare of the topsoil with high humus content. With the *humus blown out*, the moisture retention capacity of the soil is decreased. Humus regeneration is a rather protracted process.

During wind erosion, excavation landforms develop in deflation areas. Most of the soil moved is accumulated along plot boundaries where the surface is protected by vegetation (e.g. along channel banks) (Plate 5.5). The height of the resulting *accumulation landforms* following a major wind-storm can reach 0.5 m, often backfilling such ditches (Plate 5.6).

Plate 5.5 Soil accumulation along the edge of arable land at Kerekegyháza (Lóki 1984)

Plate 5.6 A ditch backfilled with soil to the west of Nyíregyháza (Lóki 1984)

The landforms in Hungary's wind-blown sand areas evolved in several phases during the end of the Pleistocene. Holocene climate, following the evolution of vegetation cover, did not favour the development of wind-blown sand landforms or the transformation of previous landforms. In areas where the original vegetation was replaced by crop cultivation, sand started to move again locally in dry periods and was deposited in variable thickness on the previous surface covering the vegetation or human settlements. This is proved by archaeological finds recovered from various depths and relatively young fossil soils. In areas of wind-blown sand, slopes are eroded (Kiss 2000) and aeolian deflation and accumulation take place at the same time. It is hard to decide which was more significant in a given area. In cultivated areas, soil eroded when thawing was blown back by winds of the arid spring period.

5.2.4 Impacts of Water Management on Agricultural Land

On arable land, a number of drainage *ditches* and other *canals* of various purposes are present. Drainage ditches are deepened where the lowest sections of the area are covered by excess water. Such ditches join in main ditches along roads. Some of them are designed for a shorter lifetime, and only left to function for years with a higher amount of rainfall; they are then backfilled or ploughed in drier years. In lowlands with a higher groundwater table, where springtime excess water represents a permanent threat to cultivation, permanent ditches are constructed in order to prevent inundation damage. The soil excavated when deepening such ditches is piled along the ditch as a positive landform ('embankment'). Maintenance of such objects is in the interest and responsibility of land owners. Natural backfilling of these ditches, a decrease in their depth is observed where no maintenance is carried out. Usually in prolonged periods of drought, maintenance is neglected.

Canals of various sizes and functions are deepened between plots of crop cultivation. In areas of irrigated rice production, water is conducted to the cultivated area through canals. Leakage of the water from the *inundated area* is counteracted by artificial dams of 50–70 cm height, which can survive long after the area is left fallow. Such dam system of uncultivated areas is present around the towns of Püspökladány and Karcag in Hungary.

In arid regions of water deficit, crop cultivation has been conducted by irrigation for ages, thus *irrigation canals* have formed the integral part of the arable landscape. In Hungary, among others in the Nagykunság Region, a range of such irrigation canals were established.

The third group of canals serves the adequate water management of arable land. The *drain pipe system* carries excess water to the canals along the plot boundaries in wetter periods whereas in dry seasons, water is supplied from the canals. Accordingly, these channels serve both water drainage and supply. From the point of view of morphology, all channels are considered to be excavation landforms.

Natural processes accelerated as arable farming created new landforms on the one hand and altered already existing natural and human-induced features on the other.

The alteration is particularly spectacular in mountains and hills. The material eroded on side-slopes alters the cross-sections of valleys (Fig. 5.3). Derasional (dry) valleys of hill regions in Hungary, where crop cultivation takes place on slopes, have undergone significant changes as a result of direct and indirect human-induced processes.

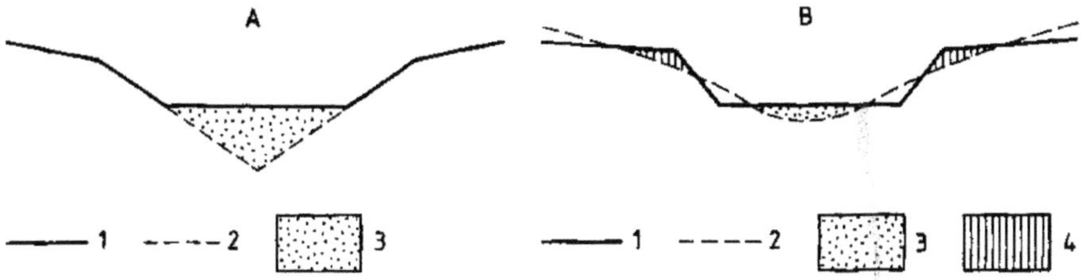

Fig. 5.3 Changes of valley cross-sections resulting from crop cultivation (**a**) Transformation of a V-shaped valley. 1 = present surface, 2 = original surface, 3 = accumulation of the eroded soil. (**b**) Transformation of a bowl-shaped valley. 1 = present surface, 2 = original surface, 3 = accumulation of the eroded soil, 4 = soil re-deposited by tillage (Figure by Hempel and Tecklenburg – after Rathjens 1979)

5.2.5 Loess Landscapes

Loess sediments cover large areas of the Earth's surface. Soils formed on loess, under favourable climatic conditions, are perfectly suitable for crop cultivation. In the Transdanubian Hills and various loess areas of the Great Hungarian Plain, human impact can be well traced. The most significant erosional landforms are produced by water; however, the rate of deflation on a windy, spring day should not be neglected either.

In addition to rainfall erosivity, soil erodibility (slope, soil moisture, soil water management and structure, surface roughness and coverage) is also very important in *water erosion*. Tillage modifies the above factors and, as a consequence, the shape of the surface. On chernozem soils formed on loess or sandy loess, erosion begins rather slowly under natural conditions. Under cultivation, soil structure and surface roughness are altered and modify water budget. As a result of precipitation – especially intensive downpours – surface erosion commences. According to Stefanovits et al. (1999) when a chernozem soil becomes 'slightly eroded', the rate of erosion will increase. It can be explained by the gradual decrease of humus and clay contents with depth. It results in the decay of resistance to water erosion. Soil degradation rate is highest when erosion reaches the parent rock (loess deposits). Obviously, surface transformation in areas with variable slope angles differs. Where the surface is not protected by vegetation, soil degradation accelerates. A good example is provided by filed paths under regular use where along with the accelerated surface degradation, *hollow roads* are formed in loess (Plate 5.7).

Wind erosion also influences soils formed on loess. In Hungary, chernozem soils of high productivity can be affected by severe deflation damage at the beginning of the vegetation period when the ploughed topsoil is not protected by vegetation. This is primarily due to the establishment of large plots exposing a large surface to the wind. Soil transported by the wind is accumulated at ditches on the edge of plots. Due to the fine granular composition, a great amount of dust is released to the air, representing a major threat to human health. As a result of wind erosion, humus content is reduced and soil erodibility increases.

Plate 5.7 Hollow road in loess (Lóki 1987)

5.3 Conclusions

As a conclusion, it can be claimed that from the environmental and soil protection point of view, one of the greatest problems of agricultural areas is *water and wind erosion,* although *mass movement* processes (landslides) also cause severe damage. On cultivated land rainwater can infiltrate deeper, while on slopes susceptible for sliding it can bring about significant changes on the surface.

Particles of the barren topsoil are removed and transported along the slope and, as a result, the topsoil is gradually thinning or quickly degrading to reduced productivity. In extreme cases, soil is eroded to bedrock and becomes unsuitable for agriculture. The extension of arable lands is also reduced by the evolution of erosion features. They gradually increase in size and depth and finally group into a gully system of several metres in depth.

Tillage causes major *changes in soil structure, porosity, ventilation and moisture conditions and temperature*. The use of heavy agricultural machinery also results in soil degradation and compaction. Soil erosion is apparent as a result of downslope tillage. In the case of improperly designed plot sizes and inappropriate technology and crop structure, wind erosion is mainly a lowland problem.

New tillage methods and extreme weather conditions seem to increase the degree of susceptibility in areas affected by soil degradation.

References

Erdősi F (1969) Az antropogén geomorfológia mint új földrajzi tudományág (Anthropogenic geomorphology as a new discipline of geography). Földr Közl 93(1): 11–26

Erdősi F (1987) A társadalom hatása a felszínre, a vizekre és az éghajlatra a Mecsek tágabb környezetében (The Impact of Society on the Surface, Water and Climate in the Broader Environment of the Mecsek Mountains). Akadémiai Kiadó, Budapest

Kiss T (2000) Futóhomok területek felszíndinamikája természeti és társadalmi hatások tükrében – dél-nyírségi vizsgálatok alapján (Surface Dynamics of Wind-blown Sand Areas with a Focus on Physical and Social Impacts – Based on Studies Carried out in the Southern Nyírség Region). Ph.D Thesis. Manuscript, DE TTK, Debrecen

Láng G (1961) Növénytermesztés (Crop Cultivation). Mezőgazdasági Kiadó, Budapest

Rathjens C (1979) Die Formung der Erdoberflache unter dem Einfluss des Menschen. Teubner, Stuttgart

Stefanovits P, Filep Gy, Füleky Gy (1999) Talajtan (Soil Science). Mezőgazda Kiadó, Budapest

Szabó J (1993) A társadalom hatása a földfelszínre – antropogén geomorfológia (The impact of society on the surface of the Earth – anthropogenic geomorphology). In: Borsy Z (ed.), Általános természetföldrajz. Nemzeti Tankönyvkiadó, Budapest

Szabó J, Dávid L (eds.) (2006) Antropogén geomorfológia (Anthropogenic Geomorphology) University notes. Kossuth Egyetemi Kiadó, Debrecen (the Hungarian version of this book)

Website Sources

http://cozay.com/documents/

http://exploringafrica.matrix.msu.edu/students/images/

http://www.kwaad.net/Photo F. Kwaad

Chapter 6
Agriculture: Grazing Lands and Other Grasslands

Csaba Tóth

Abstract Compared to industrial activities or transportation, grassland management alters the surface to a modest extent. Under favourable topographic, soil and climatic conditions, with reasonably planned grazing, the growth and consumption of grasses are in balance. When any of these factors is significantly changed, the grass cover becomes interrupted, and this influences geomorphic processes. A well-known case is the sensitive environment of the Sahel, where delicate equilibrium was upset by large-scale growth of the animal stock – aggravated by natural climate change. The extension of grazing lands and the abandonment of the traditional ways of farming induced desertification. Significant damage by trapping and soil compaction is observed at the gathering places of animals. In the arid season, tracks leading to watering holes are hardened barren surfaces. In humid periods, however, these are wetlands where intensive trampling causes the incision of ravines along the tracks. Similar phenomena can also be observed, although sporadically, under temperate climates. However, the problems of grasslands in the temperate climatic belt are also of concern of agriculture, landscape protection and nature conservation, since the subsistence of the population here is not so dependant on the utilization of grasslands as in the Sahel. Human alterations on the grasslands of Hungary took place mainly through water management, various technologies introduced to cultivation and utilization of various intensities (grazing or mowing).

Keywords Grazing · Grasslands · Desertification · Grassland management

6.1 Introduction

Agriculture, in general, especially grassland management, results in less remarkable alterations as compared to industrial activities or transportation. Apart

C. Tóth (✉)
Department of Physical Geography and Geoinformatics, University of Debrecen, Egyetem tér 1, 4032 Debrecen, Hungary
e-mail: tothcsa@delfin.unideb.hu

J. Szabó et al. (eds.), *Anthropogenic Geomorphology*,
DOI 10.1007/978-90-481-3058-0_6,

from the physical conditions, geomorphic evolution on grasslands depends on the methods and intensity of use. Under favourable topographic, soil and climatic conditions, with reasonably planned grazing, the growth and consumption of grasses are balanced. When conditions are partially or completely changed, the grass cover becomes interrupted, and this influences geomorphic processes.

This chapter intends to provide an introduction to the geomorphological problems of two grasslands under different climates through case studies. First, the causes of grass cover degradation and desertification taking place in the Sahel, i.e. in the northern marginal areas of the Sahara are discussed in detail, followed by the alterations observed on Hungary's pastures.

6.2 Anthropogenic Geomorphological Problems of the Sahel

6.2.1 Concept and Physical Geography of the Sahel

Sahel is an Arabic word meaning 'shore'. It is the name used for North Africa's sandy, dune steppe-like coastal zone, especially in Algeria and Tunisia. However, the term Sahel more often refers to the belt to the south of the Sahara, from Senegal to Sudan, along the boundary between the desert and the arid thorny savannas, between approximately 12°–15°N (Gábris 1996) (Fig. 6.1). The name Sahel was introduced to the scientific literature by Auguste Chevalier, a French botanist in the early 20th century to mark regions south of the Sahara (Élesztős et al. 1993). Due to the descending air in the Sahel, the annual amount of precipitation is low (125–250 mm); the period of rainfall lasts only for 6–12 weeks (Adams et al. 1996). There are no surface water-courses; fossil sand dunes are bound by the sparse, shrub and grass vegetation.

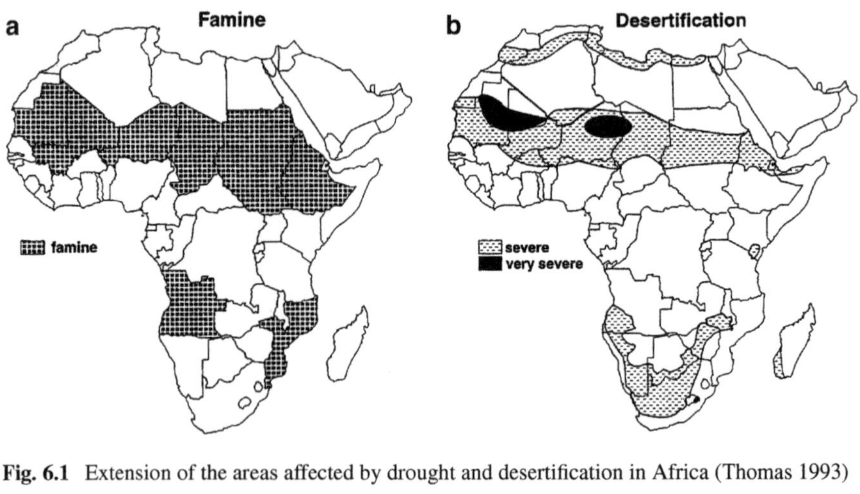

Fig. 6.1 Extension of the areas affected by drought and desertification in Africa (Thomas 1993)

The area of Sahel experienced millennia or centuries of alternating humid and hyperarid climate over the Pleistocene and Holocene. In the humid periods, the belt was characterized by humid savanna vegetation, while in the arid periods, the desert sand dunes extended far to the south (approximately to the 10°N latitude), leaving fossil sand dunes fixed by grasslands to the south of the present-day wind-blown sand areas with an annual rainfall of 1,000 mm today.

The Sahel vegetation is characterized by arid grasses, spiny shrubs and short thorny acacia trees. Soils are mostly loose sandy, while saline-clayey soils (formed on the alluvia of rivers and lakes) demand drainage. They are rather susceptible to wind and water erosion and are quickly desiccated by intensive evaporation.

In the semi-desert areas south of the Sahara, social impacts were long insignificant, and the area itself was in an ecological equilibrium. People only exploited natural resources without exhausting them. The native lifestyles and economy of peoples practising *transhumance husbandry* and subordinately (sorghum and millet) *farming* maximally adapted to the low productivity of grazing lands, and the subsistence of a limited human and animal population was ensured. The mobility of transhumance husbandry proved to be fundamental for surviving drought periods.

6.2.2 The Problem of the Sahel – Factors of Desertification

The sensitive environmental balance of the Sahel was repeatedly upset by the climate change in the past decades combined with a population boom. The resulting expansion of animal stocks together with grazing lands and the abandonment of the traditional way of farming induced desertification. As a consequence of drought years occurring since the mid-1960s, starvation and the degradation of animal stocks ensued. The dust and sand carried here by the stormy northern desert trade winds disparting the scorched surface, called the *Sahel wind,* contribute to the southward expansion of the Sahara. According to recent research, both physical and social factors contribute to desertification.

6.2.2.1 Physical Factors

First, regional climatic oscillations, reflected in altered precipitation patterns, should be mentioned. The *annual amount of precipitation* has decreased by more than 20% as compared to a more humid period between 1920 and 1939. In the more arid spell between 1965 and 1984, annual mean precipitation dropped from 200 mm to below 160 mm. This basically means the displacement of certain isohyets running roughly parallel to geographical latitudes to a southern direction to a distance of ca 50 km (Adams et al. 1996). The decrease in the amount of precipitation results in *lesser days with rainfall*, while, at the same time, the *intensity of rainfall increases*. Rainsplash erosion during major and violent downpours attacks the land surface and removes large amounts of soil. There is an increase in the *variability of the amount of precipitation* between years and rainfall becomes unpredictable. The lack of rainfall is the most common in the middle and in the end of the growing season (in August), which are the period of sorghum and millet ripening, and causes a drop in average yields.

6.2.2.2 Human Impact – Anthropogenic Geomorphologic Changes

In most of the Sahel, the low amount of precipitation allows crop cultivation only if regular irrigation is provided (Penning de Vries and Djeteyé 1982). Thus, the quantity and quality of yield highly depends on the various irrigation techniques. Recently, population increasingly concentrated next to the wells drilled and along irrigation channels. Rapid population growth is the root of environmental problems. Figure 6.2 indicates social causes in detail.

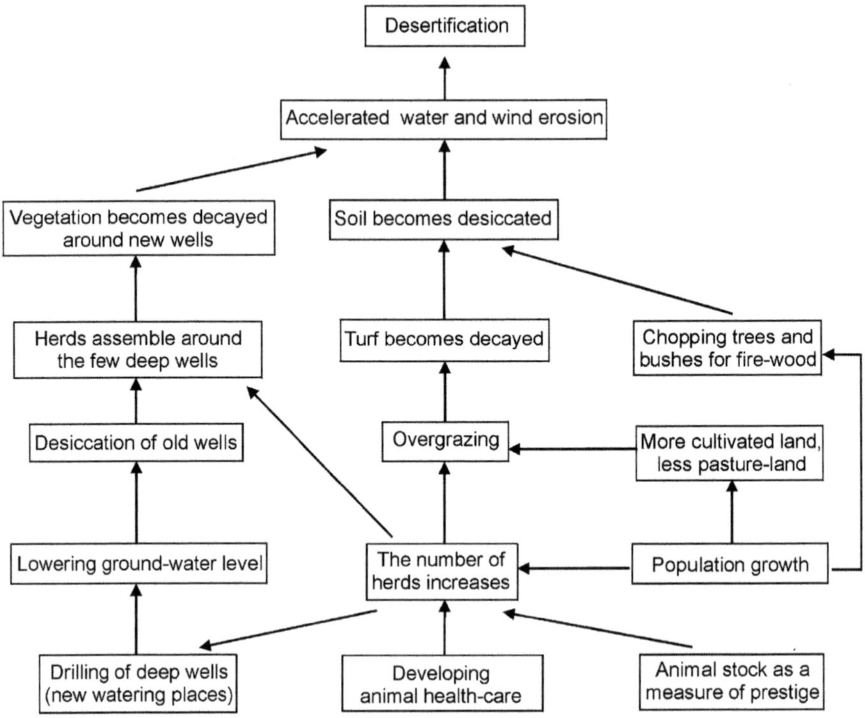

Fig. 6.2 Human factors in the desertification of the Sahel (Probáld 1996)

The fundamental problem, therefore, is not grazing but *overgrazing*, i.e. the overpopulation of animal stocks in a given area. As a result of the growth in the animal stocks induced by the population expansion, in many cases, instead of 0.5 animal unit (1 animal unit is an equivalent to 1 calf older than 2 years, or 2 so-called animals of last year younger than 2 years, or 5 sheep or goats), 5–10 animal units or even 20–25 animal units are grazed in the Sahel, accelerating the decay of grazing lands and leading to desertification (Binns 1995). Significant damage by *trapping and soil compaction* is observed at the watering holes, where animals gather. Overgrazing is detectable *in the soil profile*, in the degradation of the humus horizon. *Tracks* leading to watering holes are the sites where erosion starts. Such tracks are, in the arid season, hardened barren surfaces, while in the humid period, they are wetlands impossible to use where intensive trampling causes the incision of *ravines* along the tracks. Later a similar new track, parallel to the former one, emerges.

In the Sahel, as many as 6,000 heads of cattle as well as 50,000 sheep and goats can gather at a watering hole daily (Nir 1974). As a practice used in many countries, cattle is washed once a week in a communal tank in order to prevent diseases spread by acari. These giant tanks are found all over the region, thus the cumulative impacts of driving cattle across the land result in ravine formation and wind erosion of remarkable scale. There is a close correlation between the various levels of grazing intensity on slopes and the level of soil erosion. Overgrazing can often lead to remarkable sediment accumulation along the lower reaches of rivers.

As a result of the factors mentioned above, in the past 30 years, the total area of deserts has increased from 1.4 billion hectares to 1.6 billion hectares, an annual growth of 5–6 million hectares. Desertification, therefore, does not simply spring from the decay of the natural ecosystem but from the upset dynamic equilibrium between humankind and nature, the collapse of a traditionally well-functioning production system.

6.2.3 Solution Opportunities

In 1977, an urgent action was pronounced at the UN Conference on Desertification (UNCOD) to prevent desertification, a major problem since the tragic aridification of the Sahel belt since the 1960s. International research into desertification today studies erosion, warming (climate change – El Niño phenomena) and local circumstances blamed for such processes. Basically two kinds of solution are thought to be appropriate: governmental administrative interventions or returning to traditional farming ('indigenous strategy').

The first did not prove to be effective enough. Many of the projects related to national and international organizations failed because of resistance from the population (as, for instance, protecting the remaining arboreous vegetation by barbed wire fences) or even worsened the present situation (drilling of deep wells → greater number of animals → overgrazing). Afforestation and the widespread use of species with short vegetation period however proved to be advantageous.

As 'modern' methods did not bring a breakthrough in solving the region's problems, many researchers see the revival of traditional and conventional cultivation methods as the only way out. All agree, however, that securing social stability in the region through ending war conflicts, the solution of the Sahel's urgent problems would be substantially promoted.

6.3 Anthropogenic Geomorphological Problems of Hungary's Grasslands

6.3.1 The Distribution and Occurrence of Grasslands in Hungary

Grasslands in Hungary amount to 1,299,739 ha (12,998 km^2), of which grazing lands have a share of 871,990 ha (8,720 km^2) and meadows 427,749 ha (4,278 km^2). It means that 14% of Hungary's area is grassland (Vinczeffy 1993).

Over the past centuries, with the expansion of agriculture, the extension of grasslands has significantly dropped. By bringing high-quality grassland soils under cultivation, grasslands became almost exclusively restricted to poor-quality soils unsuitable for farming. These soils with poor water budget, inadequate nutrients supply and shallow topsoil can usually be found in the drought-stricken Great Hungarian Plain. As the average annual amount of precipitation is 574 mm and the mean annual temperature is 10.5°C in Hungary, agro-ecological surveys claim that Hungarian grasslands suffer from a rainwater deficit of 156 mm per year. It is even more unfavourable in the plains where the yearly deficit is 214 mm on average. Mountain grasslands are in a more advantageous situation as the deficit is only 51 mm. Most grasslands are found on saline soils (alluvial solonetzes and solonchaks), while others on brown forest, alluvial and bog soils (Table 6.1).

6.3.2 Grassland Classification

Permanent grasslands are classified by their utilization as (1) *pastures,* (2) *hay-fields* and (3) *meadows*. Pastures are grasslands where the fresh yields are used by grazing animals during the vegetation period. On hay-fields, grass is exclusively harvested by mowing. Meadows are mostly mown and subordinately grazed. Pastures, hay-fields and meadows are further classified by various criteria (origin, character, animal species and exposure). Based on their *origin*, pastures can be natural and artificial. Natural pasture or grasslands include original grasslands formed under natural circumstances and have avoided disruption by major human intervention. Artificial pastures, on the other hand, bear the signs of human interventions (sowing, fertilization, irrigation) along with some features of a natural grassland.

Based on their *type*, permanent and temporary grazing lands are distinguished. Permanent pastures are exclusively used for grazing. Temporary pastures are areas grazed for periods of variable length either as a necessity or for the purpose of a better use. Meadows, fallows, stubble fields, plantations and forests are examples.

Pastures can also be classified by the *animal species* which they optimally feed. From such an aspect, cattle, horse, sheep, goat and goose pastures are distinguished. For *cattle pastures*, lowlands well covered by good-quality fibrous long grasses are the most suitable. Cattle spare most of the grass cover as they pick grass by their tongue, leaving adequately high, ca 2 cm stumps. Total mastication only occurs when the pasture is overgrazed. *Horse pastures* are areas with harder soils and

Table 6.1 Main features of the soils of Hungary's grasslands (Vinczeffy 1993)

	Area			Area			Area	
Main types	1,000 ha	%	Slope gradient (%)	1,000 ha	%	Humus content	1,000 ha	%
Skeletal soils	84	7	0–5	650	50	0	40	3
Brown forest soils	238	18	5–15	290	23	0–1	125	10
Chernozem	130	10	15–25	280	22	1–5	694	54
Saline soils	392	31	Above 25	63	5	5–10	272	21
Meadow soils	217	17	Total	1,283	100	Above 10	152	12
Bog soils	161	12	pH			Total	1,283	100
Other soils	61	5	3.5–5.5	156	12	Soil water management according to water intake and storage		
Total	1,283	100	5.5–6.5	640	50			
Soil depth			6.5–7.5	276	21	Very poor	490	38
			7.5–8.5	87	7	Poor	310	24
2–20 cm	503	39	Above 8.5	9	1	Moderate	186	15
20–50 cm	374	29	Mixed	115	9	Good	177	14
>50 cm	406	32	Total	1,283	100	Excellent	120	9
Total	1,283	100				Total	1,283	100

harder grass. Horses snap at greater depths compared to cattle picking the grass with their lips. They often cut grassland vegetation and also trample heavier and, as a consequence, exploit pastures to a greater degree. *Sheep* and *goats* exploit the short-growing, less valuable turf of arid areas. Sheep snap at greater depths, often getting hold even of stern nodes of grasses and pulling up the whole plant from the soil while grazing. Goats tend to consume all vegetable parts (including the shoot and seedlings of low bushes) leaving barren lands behind. Sheep and goat pastures in Hungary are mostly arid areas of low fertility. Therefore, valuable plant stocks can only be maintained on them by regular management (including fertilization). When *pigs* are unable to find an adequate amount of forage, they dig up the whole pasture. Adequate care should be taken, otherwise the whole grassland can be easily destroyed. Among the domestic fowls, *geese* are the greatest hazard to pastures. Therefore, grazing areas of the poorest quality should be designated for them. Geese consume all green vegetable parts and pollute the air by their feathers and caustic excrements. Stocks of several hundreds leave barren surfaces behind on which only some aggressive weed species are able to grow.

Based on *exposure*, mountain, hill, lowland, riparian and wetland pastures are distinguished. Mountain pastures are situated at an elevation of 800–1000 m, having abundant and substantial vegetation. Hill and lowland pastures are often eroded, mostly in arid locations and grasses have a moderate growth rate on them. Riparian pastures are found along rivers and covered by grasses. They are among the most valuable pastures, particularly suitable for cattle – if not waterlogged. Wetland pastures in waterlogged areas usually have a harsh vegetation mainly consisting of sedges. They are mainly utilized by pigs and, to a limited degree, by cattle.

6.3.3 Geomorphological Consequences of Grassland Management

In order to achieve a better exploitation of pastures and to maximize grass yields, humans have been intervening in the natural development of grasslands for centuries (Baksay 1996). In Hungary, especially in the Great Plain, nomadic herding was common until the late 13th century. In the 14th century, permanent settlement and the first regulations on grazing marked the advent of grassland management. Grassland management in the modern sense, however, did not emerge until the late 19th century and began to develop dynamically after World War II.

Human interventions affecting grasslands can be divided into four groups: (1) water regulation of grasslands, (2) various cultivation techniques, (3) fertilization and (4) utilization (grazing, mowing). The latter two factors have already been discussed and, therefore, the geomorphologic consequences of water regulation of grasslands and cultivation techniques are reviewed here.

6.3.3.1 Water Regulation of Grasslands

Abundant grass yields require water regulation. It has a twofold purpose: on the one hand, harmful excess water has to be removed; whereas on the other, an adequate amount of water must be supplied for the grasses at the right date. The most common and, geomorphologically, the most spectacular way of fighting waterlogging is *open-ditched land drainage*. Excess water is essentially drained by a network of lateral and collecting ditches interconnected from waterlogged areas (Plate 6.1). Lateral ditches join the wider and deeper trunk ditch which allows constant water flow. The density of the ditch network depends on the soil cohesion and the degree of waterlogging. A disadvantage is that the ditches occupy extensive areas, impede transportation and large-scale cultivation, the annual costs of cleaning is not negligible either. However, it has the advantage of draining large amounts of water within a short period of time at a relatively low cost.

Plate 6.1 Drainage ditch on a saline pasture in the Hortobágy (Tóth 2003)

On cohesive soils with appropriate stability, *uncased mole draining* made by drainage-plough is also applied. Underground hollows can easily cave in on sandy soils and bog soils, thus the technique cannot be applied there. Compared to the use of plastic drain pipes, they are cheaper, however with a shorter duration.

By draining excess water, significant changes can take place in the grass associations. By dropping groundwater table, acidic sedges and other plants (Cyperaceae, Yellow Iris – *Iris pseudacorus*, Ranunculaceae, Water Mint – *Mentha aquatica*, Pennyroyal – *Mentha pulegium*) decrease, with them having been replaced by sweet grasses (Creeping Bent – *Agrostis stolonifera*, Meadow Foxtail – *Alopecurus pratensis*, Reed Canary-grass – *Phalaris arundinacea*).

Based on experiments carried out in Hungary, it can be claimed that the undisturbed evolution of turf mixtures requires an annual amount of precipitation of approximately 700 mm under the climate of the country. On the contrary, however, grasslands in lowlands only receive 548 mm rainwater on the average. Consequently, water supply must also be provided. Four types of *irrigation* are known: (1) *flooding,* (2) *border dyke irrigation,* (3) *surface sprinkler irrigation* and (4) *'nesting' and furrowing.* Of the four methods mentioned above, 'nesting' and furrowing cause spectacular geomorphologic alterations on pastures. 'Nesting' and furrowing cannot actually be regarded as irrigation as they only focus on the retention of winter snow-melt and spring rainwater. During nesting, depending on the gradient of the field, ridges 20–30 cm in height should be formed by commonly applied ploughs with a distance of 60–100 m along the contour lines, then perpendicular to them. The network of rectangular nests impedes the immediate runoff of snow-melt and early-spring rainwater by retaining the adequate amount of water. Therefore, there is no need for irrigation proper.

The deepening of drainage and irrigation ditches, especially in saline areas can trigger significant erosion processes. On channel banks with more disturbed soil, landforms similar to natural salt berms; (however, developing more quickly and being more remarkable) are observed. *Salt berm erosion* is common here. Above the lower, older and steeper berm edge, the top berm is intensively eroded by rilling and sheet wash, and gradually decays into a barren and glaring white surface, where some specimens of halophilous specialist plant species appear. The intensive lowering of the lower berm top brings about an upper, younger and more gently sloped berm edge (Plate 6.2). Incising into the berm surface, the saltwater rills create gaps across the berm edge. In the foreground of this hollow, the rills usually accumulate miniature alluvial fans. The intensive decay of berm tops often produces surfaces similar to badlands on which percolating water often springs along berm forefronts (Plates 6.2 and 6.3). This process requires a relatively steep gradient, provided along the banks of artificial drainage and irrigation ditches (2.5 m drop at 10 m distance). In addition, intensive trampling by animals (Plate 6.3) and heavy showers also play

Plate 6.2 Formation of a two-stepped salt berm on the edge of a drainage channel in the Hortobágy (Tóth 2003)

Plate 6.3 Channel bank intensively formed by rill erosion and sheet wash in the Hortobágy (Tóth 2003)

a major part in berm formation. Runoff from showers can easily dissect disturbed, loose ditch banks. Under favourable conditions for erosion, annual berm regression can reach 10–15 cm and a surface lowering of 2–3 cm is observed, resulting in a gradual decrease of pasture area year by year (Tóth 2003).

6.3.3.2 Cultivation Techniques of Grasslands

Stony pastures frequently occur in mountain areas and river valleys of Hungary. Unfortunately, more and more pastures accommodate immense amounts of rubble, bricks, glass, tiles, etc. Stones are harmful on pasture as they cause injuries to animals, also dehydrating the soil when intensively warmed and spreading weeds. Therefore, the *collecting of stones* on the surface and at low depths is an important anthropogenic intervention on pastures. The stones collected are piled around watering holes and wells, and used for the construction of dams to control gullies, marking borders, road construction and other purposes.

On abandoned grasslands densely overgrown by bushes, a significant amount of area is occupied by shrubs instead of grasses. *Shrubs* have to be *cut* to improve conditions.

Overgrazing and the *damage* caused by *power machines* (wheel tracks) usually result in stripping of the closed grass cover, eventually the decay of soil structure (Plates 6.4 and 6.5). When the mass of grass cover amounts to 180 g/m^2 in dry mass, it provides excellent protection against erosion (Vinczeffy 1993). A natural consequence of the stripped surface and neglected soil scars, which mainly occur in mountainous and hill areas, the grass cover is interrupted and removed by

Plate 6.4 Grass cover disrupted by intensive grazing (of sheep) in the Hortobágy (Tóth 2003)

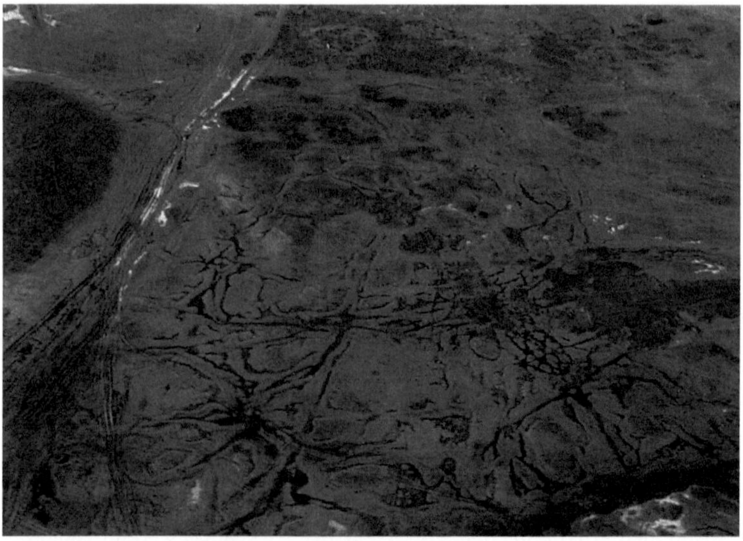

Plate 6.5 Saline grassland disrupted by heavy trucks in the Hortobágy (Tóth 2003)

incising gullies after heavy showers. The embryonic gully is broadened and deepened by runoff water. The gully banks cave in and extensive slumping takes place. The soil decayed will be transported by runoff water, resulting in further undercutting. Consequently, widening and deepening *gullies, ravines and gorges* are formed. Further development of steep gullies can be hindered by reducing slope along some bank sections, bringing them to trough shape, constructing dams of stones, stakes and twigs and planting *Pseudoacacia* trees and grasses.

When sand grasslands are overgrazed, not only erosion but also *deflation* becomes more intensive. Deflation can also be controlled by grassing.

In unmanaged grasslands on good soils, *mole-hills* are common. With time they are overgrown by grasses and colonized by ants. Thus, from a small mole-hill an ant-hill even half a metre high can develop in the course of time. When 15–20% of the grazing land in question is occupied by mole-hills and ant-hills, they are commonly levelled by a hoe or cleared away by shredders and planishers.

Tussocks also contribute to the unevenness of grazing lands. They partly develop by the growth of the roots of *Carex* species (e.g. Tussock sedge – *Carex stricta*); on the other hand, clumps of various sizes can be torn out by the trampling animals grazing on soggy soils. During intensive grassland management such small rises are pressed in the soil by plain radial roll as well as grass-comb.

6.4 Summary

Until transhumance husbandry along the desert margins or nomadic grazing under temperate climate was the only way of utilizing grasslands, equilibrium conditions were maintained between nature and human society. Especially, the changes in social conditions (population growth, the increase of animal stock, intensive grassland management, overgrazing) can result in the imbalance of grass ecosystems. This is well indicated by the example of the Sahel, an environmentally sensitive area. However, similar phenomena can also be observed, even if sporadically, under temperate climate. Solving the urgent problems of the Sahel, survival and improved quality of life can be ensured for millions of people. Although the subsistence of the population in the temperate climatic belt is not so dependent on the utilization of grasslands as in the Sahel, the problems of grasslands are also of concern to agriculture, landscape protection and nature conservation. Water management, various technologies introduced to cultivation and utilization of various intensities (grazing or mowing) have substantially altered the conditions of the grasslands of Hungary.

References

Adams WM, Goudie AS, Orme AR (eds.) (1996) The Physical Geography of Africa. Oxford University Press, Oxford

Baksay TB (1966) Legelő és rétművelés (Pasture-land and Meadow Cultivation). Mezőgazdasági Kiadó, Budapest

Binns T (ed.) (1995) People and Environment in Africa. John Wiley and Sons Ltd, England

Élesztős L et al. (ed.) (1993) Magyar Nagylexikon 16 (The Encyclopedia of Hungary Vol. 16). Száhel. Magyar Nagylexikon Kiadó, Budapest

Gábris Gy (1996) Szudán: átmeneti öv a Szaharától délre (Sudan: a transitionary zone to the South of the Sahara). In: Probáld F (ed.), Afrika és a Közel-Kelet földrajza. ELTE Eötvös Kiadó, Budapest

Nir D (1974) Man, A Geomorphological Agent. Keter Publishing House, Jerusalem
Penning de Vries FWT, Djeteyé MA (eds.) (1982) La productivitédes des paturages Sahéliens: Une etude de sols, des végétatioins et de l'exploitation de cette resource naturelle. Wageningen, Neth
Probáld F (ed.) (1996) Afrika és a Közel-Kelet földrajza (The Geography of Africa and the Near East). ELTE Eötvös Kiadó, Budapest
Thomas DG (1993) Sandstorm in a teacup? Understanding desertification. Geogr J 159 (3): 318–331
Tóth C (2003) A hortobágy negyedidőszak végi felszínfejlődésének főbb természeti és antropogén vonásai (Main Physical and Anthropogenic Features of the Landscape Evolution of the Hortobágy at the End of the Quaternary). (Ph.D.) Dissertation. Debreceni Egyetem, Debrecen
Vinczeffy I (ed.) (1993) Legelő és gyepgazdálkodás (Pasture-land and Grassland Management). Mezőgazda Kiadó, Budapest

Chapter 7
Agriculture: Cultivation on Slopes

Péter Csorba

Abstract Among all anthropogenic geomorphological features, agricultural terraces on slopes have the largest relief-modifying impact, equally affecting soil, climatic, hydrologic and biogeographic conditions. Terraces are the predominant elements of landscapes in many parts of the world. The techniques of terrace construction and maintenance show regional characteristics, adjusted to the local natural resources, economic demands or crop requirements. The size of terraces primarily depends on slope inclination, their material on the availability of stones and their degradation on climate. Staircase terraces are an obstacle to air movements and, therefore, have a significant microclimatic impact. While in Hungary terraces are applied almost exclusively in vine plantations, in the Mediterranean, in addition to vineyards, terraced olive, citrus and chestnut plantations are almost equally common. The maintenance and reconstruction of agricultural terraces is an important task in landscape conservation, particularly at World Heritage sites.

Keywords Slopes · Cultivation · Erosion · Terraces · Vineyards

Among land-use types, agriculture demands the largest area. Humans have always intended to use lands of moderate slope gradient for arable land and plantations. Steep slopes susceptible for landslides, rock-falls and collapse were only brought into long-term use when there was no other suitable land or a valuable product that was worth the high risk and surplus investment with significant costs could not be produced. Too high a relief is a major obstacle to cultivation all over the world – although in various degrees on the different continents.

According to the data in Table 7.1, Europe is, with the exception of nutrient deficiency, in a far more advantageous position for all indicators.

There is an urgent need for the utilization of steep hillsides in major population concentrations, i.e. in Southeastern Asia from Pakistan to Japan. A similar

P. Csorba (✉)
Department of Landscape Protection and Environmental Geography, University of Debrecen, Egyetem tér 1, 4032 Debrecen, Hungary
e-mail: csorbap@delfin.unideb.hu

DOI 10.1007/978-90-481-3058-0_7, © Springer Science+Business Media B.V. 2010

Table 7.1 Percentages of limiting soil properties for plant growth by continent (after Rid 1984)

	Relief	Nutrient deficiency	Aridity	Extreme wetness	Permafrost	No significant limitation
Europe	12	33	8	8	3	36
North and Central Asia	38	9	17	13	13	10
South Asia	43	5	23	11	–	18
Africa	13	18	44	9	–	16
North America	10	22	20	10	16	22
Central and South America	14	31	25	10	–	20
Australia	8	6	55	16	–	15
World	22	23	28	10	6	11

demographical pressure is apparent in Central America, whereas, as shown by the table, agriculture in North America, Africa and Australia is less restricted by such topographic disadvantages.

Topographic change associated with agriculture is less significant in the case of *grazing*. Where on 25–30% slopes, large-scale landscaping (the establishment of terraces) is required for crop cultivation and plantation purposes, there are no notable topographic obstacles to grazing. As large animals are able to graze when standing in slope direction, erosion by animal treading along contour lines will create terracettes, regular 'visible contour lines' ('cow isohypse'; Plate 7.1).

To control erosion during the *afforestation* of steep slopes, smaller scarps are made to dissect the slope. Such features are gradually washed away within 10–20 years. For forest management and transport, roads are cut into slopes of a gradient exceeding 25–30% and mass movements often result. Deforestation in the Mediterranean is accompanied by interventions with severe geomorphologic consequences. Since the 1970s many forests have been destroyed in summer fire events. It is understandable, therefore, that forests burned down in the surroundings of famous holiday resorts had to be restored most urgently – even for landscape aesthetic reasons (Plate 7.2). As a solution providing both profit and green landscape, the plantation of eucalyptus trees is preferred in countries like Portugal, where this species already composes 30–35% of the forest stands (Horváth et al. 2003). Most of the eucalyptus plantations for industrial purposes are placed on terraced slopes, creating a rather regulated, uniform and artificial landscape.

From the point of view of geomorphology, the impact of land-use types mentioned above is negligible compared to *large-scale terracing*. Among the anthropogenic geomorphologic consequences of agriculture and forest management, only the construction of artificial terraces on slopes has a relief-modifying impact that *affects all landscape factors* in an area, including soil, climatic and hydrologic as well as biogeographic conditions. Terraces become the predominant element of landscape functioning and scenery and can even be regarded as a new landscape type (Csorba and Zsadányi 2003).

Plate 7.1 'Visible contour-lines' trod by grazing animals near Eger-Síkfőkút (Csorba 1974)

Plate 7.2 Afforestation following a forest fire in South Portugal (Csorba 2005)

Terraces are constructed in Hungary almost exclusively for vine plantations (Csorba and Novák 2003; Lóczy and Nyizsalovszki 2005). In the past decades, arable terraces have been abandoned almost everywhere in the country. In the Mediterranean, in addition to vineyards, terraced olive, citrus and chestnut plantations are almost equally significant. It is more characteristic, though, that rice being a staple in tropical and subtropical regions is grown on extensive terrace systems. It is also usual that coffee, cocoa, pepper and tea plantations occupy terraced slopes in tropical landscapes. Strip cultivation of steep slopes in Southeastern Asia goes back to thousands of years. Moreover, in this region, strip cultivation agriculture also has to overcome violent tropical rainfalls and earthquakes.

The *establishment of terraces* starts with landscaping. Taking advantage of the natural unevenness of slopes, terraces of various heights, widths and materials are constructed. This agrotechnical solution was applied usually for slopes steeper than 12–17%. Strip cultivation is rather expensive with a significant (10–20%) net area loss occupied by terrace risers. Terracing obviously obliterates former flora and fauna entirely, topsoiling is usually necessary and water regime is also considerably transformed – particularly in the case of irrigated strip cultivation, which inevitably involves the drainage of terrace surfaces.

The technique of terrace construction and maintenance shows regional characteristics, although these are adjusted to the local natural resources, economic demands or crop requirements, for instance, whether there are suitable stones available nearby to build supporting walls, what should be the capacity of transportation roads between terraces, how large should be the spaces and distances between stocks and rows. The first authentic documents on the construction of rock terrace supports around Tokaj originate from the 1620s. Thus, at some locations, the terraces and stone walls visible today were constructed 250–300 years ago (Boros 1996; Plate 7.3).

Plate 7.3 Old stone walls near Mád (Tokaj-Hegyalja Region) (Csorba 2006)

Where stones appear on the surface, they are frequently collected and piled up along plot boundaries. The coarse debris will, in the course of time, make up a dike along the slope with a length of 50–100 m, being a characteristic anthropogenic topographic element, e.g. of vineyards in the vicinity of Tokaj. These downslope stone stripes with vegetation cover only along their rims are locally called 'obala(s)'. They have a typical microclimate and, remaining relatively undisturbed, they provide valuable ecological shelters for biota.

The size of terraces primarily depends on slope inclination. The steeper the original slope is, the narrower the terraces can be constructed with higher risers. Exact size data show a high deviation, as there are staircase terraces with a width of 2–3 m as well as extensive terrace surfaces of 40–60 m width, which also allow mechanical cultivation.

The next basic difference is whether the *terrace tread* is:

- *horizontal*,
- *sloping at 4–8%* or
- *of reverse gradient*, i.e. sloping inward by 3–5%.

The slope of terrace treads is usually 4–8%. On terrace surfaces of lesser gradient, debris accumulation is common and produces various anthropogenic geomorphologic landforms (Plate 7.4).

Plate 7.4 Loess mud accumulated after a violent summer shower on a vineyard terrace near the village of Tarcal (Tokaj-Hegyalja Region) (Csorba 1996)

Terrace risers can be

- *slopes of earth*, mostly grassed, or
- *supporting walls*, extremely steep or subvertical walls of stone or concrete.

The solution to be applied is usually selected by the potential erosion hazard (depending on the erosivity of precipitation and the erodibility of the terrace surface). Terracing undoubtedly improves moisture storage capacity, effectively hinders erosion, impedes the removal of fertile soil or (chemical) fertilizers, facilitates irrigation and mechanized cultivation. On loess, however, a horizontal or inward terrace tread may result in disastrous consequences (as for the 'Rákóczy-szőlő', a Tokaj vineyard terraced in 1960–1961). Due to the high vertical permeability of loess, water remaining on the surface or assembled in the inward inclined section of the slope led to extensive dissolution, piping, within a few years (Fig. 7.1). It is inevitably triggered by the ponding of rainwater in a terrace corner, dissolving calcium and concentrating chloride. This chemical process is then followed by the development of increasingly widening drainage routes, inducing mechanical loess degradation. Stucking a pole and concrete pillars into the terrace surface, the loess structure was destructed and piping intensified (Plate 7.5). Within a few years, as slumps of 2–3 m both in diameter and depth developed on the terrace surface, viniculture had to be abolished. In order to prevent damage, prolonged waterlogging has to be precluded and terrace surface slope has to have a 1–2% slope. It is a further advantage when vine poles and pillars are placed on a low ridge making water unable to dissolve passages along them.

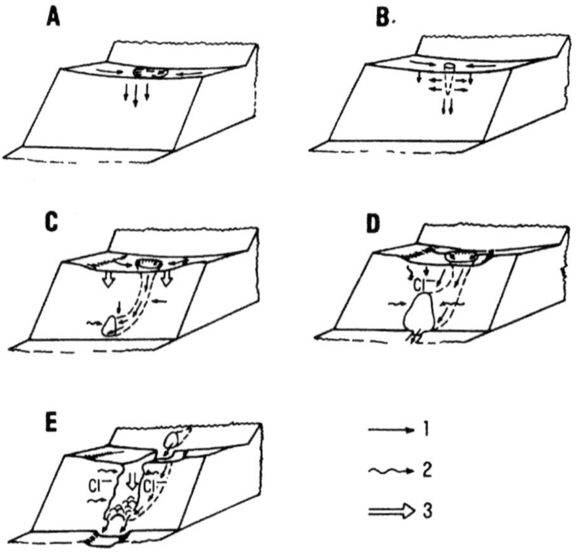

Fig. 7.1 Loess degradation through piping (Kerényi and Kocsis-Hodosi 1990)

Certain techniques have always been available to *drain rainwater*. Among them, the most simple was when settling pits (so-called lector pits or lectors) were constructed along the margins of all terraces, and runoff from the plots and the terrace surfaces were collected in them. This way, the ponding of surplus rainwater, representing an erosion hazard increasing towards the footslope, was counteracted.

Plate 7.5 Loess degradation through piping. Underground passages formed by impounded water around a vine pole on a terrace surface (top left) in the course of time, conduct water to the lower terrace (centre), Nagy Kopasz Hill, Tokaj (Csorba 1988)

The digging of lector pits was already ordered by a regulation on vine cultivation of Tokaj from 1641.

Today, concrete trench drains are constructed adjacent to major terrace systems, which surround the entire terraced slope and effectively drain the water received from nearby areas. Similarly to other European vine-growing regions, V-shaped concrete paved transport roads also function as drainage ditches (Plate 7.6).

Plate 7.6 A concrete road also functioning as a drainage ditch between terraced irrigated vineyard plots, in the vicinity of Tarcal (Tokaj-Hegyalja Region) (Csorba 1985)

In order to mitigate erosion damage from surface runoff, the *protection of slopes* cannot be neglected. Where terrace walls, 3–8 m high and inclining at 70–90%, are composed of the local material, they are usually densely grassed. In Hungary, the plantation of Robinia trees is a common means to stabilize slopes (in the case of the high bank along the northeastern shore of Lake Balaton or around the village of Pere in the valley of the River Hernád, Northeastern Hungary). At other locations, strawberry plantations and fruit trees were also planted (or preserved) on terrace risers. The grass cover involves an ecological hazard: it may shelter pests and, therefore, a thorough chemical weed control may be necessary. Covering slopes with plastic foils is meant to tackle the problems caused by both erosion and weeding. Foil cover, indeed, also reduces the loss of humidity by evaporation and, thus, promotes savings on irrigation. This technique is not yet widely used as foils are easily damaged, decomposed by sunshine in 2–3 years; the types resistant to weather and mechanical damage are rather expensive. (The terrace slope seen left on Plate 7.6 is covered by an experimental foil strip.)

Among the disadvantageous features of terraces, primarily the *adverse changes in the microclimate* are usually emphasized (Csorba 2006; Nyizsalovszki and Lóczy 2008). Staircase terraces are an obstacle to air movements. As a consequence, cold air reaching the terrace will not be able to move further towards the slope, forms a permanent air pond and susceptibility for frost in summer and autumn significantly increases (Justyák and Pinczés 1976). As proved by a *frost* damage mapping project carried out at Tokaj (Pinczés and Marton-Erdős 1983), strip cultivation resulted in severe damage in vineyards where previously such natural disasters occurred only rarely. The zone susceptible to frost stretches to an elevation 10–20 m lower, and this threat has become more common at the midslope, with rather favourable ecological conditions in other respects. Frost damage mapping also proved that the degree of frost damage decreased towards the outer margins of terraces (Fig. 7.2). The most intensive frost damage could be located to the inward base of terraces exceeding 10 m in width, horizontal or with reverse gradient (Pinczés and Marton-Erdős 1983). At the foot of strip-cultivated slopes, public roads or railway lines are often found, the embankments of which frequently impound cold air flow and result in frost damage on lower terraces.

The impoundment of air flow by strip cultivation does not only increases frost hazard, but also increases the temperature of *pockets of hot air* during the summer. This increases evaporation demand and cultivated crops use more water. In the wine-producing regions of Germany, Austria and France under a more humid climate, the tendency to mist and, consequently, the hazard of fungoid diseases is higher in the valleys between terraced slopes. The modified microclimate of terraces will influence soil temperature, humidity balance, indirectly influencing the intensity of chemical processes and microbial activities in soils. It is detected that humidity conditions decrease the value of irradiation, which influences the quality of yield of crops with higher demand of sunlight.

For downslope-cordoned vine plantations on terraces, variations in irradiation between the outward and inward sides are less developed and the hazard of fungoid

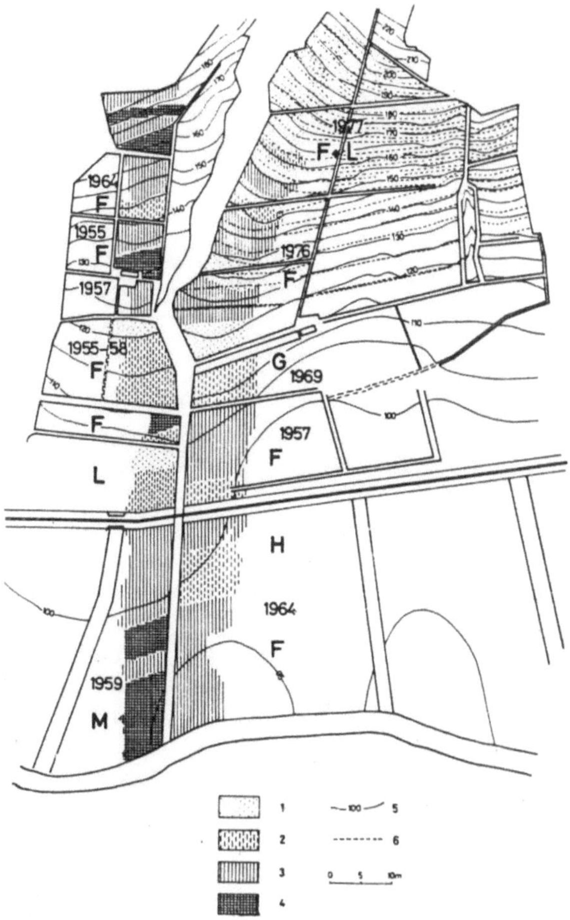

Fig. 7.2 Frost damage map for the vineyard terraces of the Szarvas Farm Road at Tarcal. 1: slight; 2: moderate; 3: severe; 4: extremely severe frost damage, 5: contour-line; 6: terrace margin. Years indicate the time of vine plantation. (Pinczés and Marton-Erdős 1983)

disease is higher in the inward, shaded, side, cooler and more humid, where the sugar contents of berries will also be lower.

Terraces, especially systems supported by stone walls – as indicated by one of the examples above – are rather *permanent features* in the landscape. Despite occupying only a fraction of the area, they have a central role in the protection of some landscapes. Landscape units renowned for strip cultivation are often emblematic regions of countries, like Tokaj or Badacsony (on the northern shore of Lake Balaton) in Hungary. In addition, strip cultivation takes place on foothills; terraces are visible from long distances. It is probably not a coincidence that a number of cultural *World Heritage* landscapes present strip cultivation (see http://whc.unesco.org/en/list World Heritages Sites). The most spectacular among them are the rice terraces in the Philippines and in Java, but the vineyards along the River Duoro in Portugal, Cinque Terre in Italy (Plate 7.7) or the agricultural

Plate 7.7 Vineyard terraces on steep slopes facing the Ligurian Sea (Cinque Terre, Italy) (Csorba 2001)

landscape of the Wachau Gorge in the Austrian section of the River Danube are also unique landscapes.

There are probably inordinately 'over-regulated', almost entirely *reconstructed landscapes*, as e.g. Kaiserstuhl in the Rhine Valley (Plate 7.8), or the wine region facing Lake Biel in Switzerland and the Mosel Valley in Germany. On the other hand, *abandoned* and declining *terrace systems* are not visually attractive.

Plate 7.8 800 km of vineyard terraces were reconstructed in the Rhine Valley, on the Kaiserstuhl Hill (Germany) between 1969 and 1972 (Csorba 2000)

Uncultivated terraces retain their intense landscape-forming impact for decades; secondary vegetation on abandoned terraces will not mask the physical, ecological and visual impacts of the former cultural landscape, as at several locations in the Rhine Valley. The question arises: what to be done with crumbling stone walls? Their restoration or reconstruction is rather expensive. Strip cultivation has a strong and long-term impact on the present and future land-use potential of slopes. When terraces are built, land use becomes fixed and more flexible land-use practices are excluded.

The construction of terrace systems is cost-intensive and demands massive efforts from the community. Their *maintenance* also requires permanent labour investment. Therefore, a worldwide decline in rural population, especially in its active groups capable of work, means that less and less people are left to maintain the terrace and irrigation systems. It is evident that the shortage of labour in the countryside will soon be reflected in the conditions of these remarkable structures. For some time, the abandonment of terraces in peripheral location would not prohibit the functioning of the system; however, when the maintenance of crucial points of the system is neglected, landscape functioning will be severely inhibited.

References

Boros L (1996) Tokaj-Hegyalja szőlő- és borgazdaságának földrajzi alapjai és jellemzői (Geographical basis and features of the vine and wine cultivation of the Tokaj-Hegyalja Region). Észak- és Kelet-Magyarországi Földrajzi Évkönyv 3. Főiskola Földrajz Tanszéke, Nyíregyháza

Csorba P (2006) Egy bodrogkeresztúri hóolvadás menetének tájföldrajzi elemzése (A landscape geographical analysis of the course of thaw at Bodrogkeresztúr). In: Csorba P (ed.), Egy szakmai életút eredményei és helyszínei. Tiszteletkötet Martonné Dr. Erdős Katalin 60. születésnapjára Debreceni Egyetem, Debrecen

Csorba P, Novák T (2003) Veränderungen der Landschaftsstruktur und Landnutzung in Ungarn nach dem EU-Beitritt. In: Bastian O, Grunewald K, Schanze J, Syrba RU, Walz U (Eds.) Bewertung und Entwicklung der Landschaft, Ergebnisse der Jahrestagung IALE-Deutschland 2002 in Dresden. IÖR–Schriften, Band 40, Leibniz-Institut für ökologische Raumentwicklung, Dresden

Csorba P, Zsadányi E (2003) Veränderungen des Characteristikums der Naturnahen Vegetation in den Letzten 30 Jahren in dem Halbbecken von Bodrogkeresztúr. Acta Geogr Debrecina 36: 121–140

Horváth G, Makádi M, Munkácsi B (2003) A táj és a környezet változásai Portugáliában és Magyarországon (Changes in the landscape and environment in Portugal and Hungary). In: Csorba P (ed.), Környezetvédelmi Mozaikok, Tiszteletkötet Dr. Kerényi Attila 60. születésnapjára. Debrecen Egyetem, Debrecen

Justyák J, Pinczés Z (1976) A domborzat fagykármódosító hatása Tokajhegyalján (The relief's frost damage modifying impact in the Tokaj-Hegyalja Region). Földr Ért 1: 31–60

Kerényi A, Kocsis-Hodosi E (1990) Löszpusztulási formák és folyamatok kvantitatív vizsgálata szőlőterületeken (Quantitative analysis of loess degradation landforms and processes at vine-growing regions). Földr Ért 1–4: 29–54

Lóczy D, Nyizsalovszki R (2005) Borvidékeink földhasználat-változásainak tájökológiai értékelése (Landscape ecological evaluation of the changes in land-use in Hungary's wine regions). Tájökol Lapok 3 (2): 243–252

Nyizsalovszki R, Lóczy D (2008) Tradicionális tájhasználat és terroir Tokaj-Hegyalján (Traditional land-use and terroir in the Tokaj-Hegyalja Region). In: Csorba P, Fazekas I (eds.), Tájkutatás, tájökológia (Landscape research, landscape ecology). Medián Foundation, Debrecen 111–116

Pinczés Z, Marton-Erdős K (1983) Untersuchung eines Winterfrostschadens in der Tokajer Weingegend. Acta Geogr Debrecina 21: 25–38

Rid H (1984) Das Buch vom Boden. Ulmer Verlag, Stuttgart

Website Source

http://whc.unesco.org/en/list World Heritages Sites

Chapter 8
Agriculture: Deforestation

Zoltán Karancsi

Abstract One of the most spectacular environmental changes caused by human activities results from forest clearing. The destruction of forests in the Mediterranean started in Europe about 2,000–3,000 years ago, along with the development of transportation (shipbuilding, the appearance of wheeled carts). By the 20th century, deforestation became focused on the tropics and today the destruction of the tropical rain forests takes place at a rate never witnessed before. The reduced water-retaining capacity (interception) of the foliage of rainforests will have serious consequences. Intense rainfalls can entirely wash away the laterite soils from the bedrock and make revegetation impossible. A historical overview is provided for a study area in Hungary, which has been used for agriculture and forestry over several millennia. The impact of clear-cutting on runoff (flash floods) is proved by evidence. The spread of *Robinia pseudoacacia* stands is a major environmental problem.

Keywords Forest cover · Deforestation · Water budget · Soil erosion · Medves region · Forest management

8.1 Introduction: The Importance of Forests

Today forests cover only 26% of all continents (38.7 million square kilometres). This land-use type has the highest organic matter production capacity. The total amount of forest organic matter is estimated to be 950 billion tonnes. Forests, including tropical rainforests, temperate deciduous forests and boreal coniferous forests, present the highest diversity in their climatic belts.

The *positive impacts of forests* are mainly manifested by their soil conservation function. Foliage, litter, undergrowth and roots retain a significant amount of water,

Z. Karancsi (✉)
Department of Geography and Ecotourism, Institute of Applied Science, Faculty of Education, University of Szeged, Boldogasszony sgt. 6, 6725 Szeged, Hungary
e-mail: karancsi@jgypk.u-szeged.hu

J. Szabó et al. (eds.), *Anthropogenic Geomorphology*,
DOI 10.1007/978-90-481-3058-0_8, © Springer Science+Business Media B.V. 2010

reduce the rate of surface runoff, and in the meantime increase infiltration. (Leaf-litter covering the surface has a high water-retaining capacity. The infiltration of water into forest soils is 7–8 times higher, surface runoff 2–3 times higher in forested surfaces compared to areas not covered by forests.) The age of forests can also impact *water budget* in various ways. The older the forest stand is, the more surface runoff it is capable to retain (Illés and Konecsny 2000). In addition, forests are habitats of a number of plant and animal species, thus unreasonable deforestation is most dangerous to the biosphere. Apart from their genetic values, forested areas are also valuable from the point of view of the adequate *functioning of the Earth's system*. Forest stands have an important role in the exchange of CO_2 and O_2 gases. Trees incorporate a significant amount of CO_2 into their bodies. (Each square metre of the tropical rain forests absorbs about 1 kg coal from the atmosphere, i.e. 10 t/ha (Kerényi 2003).) With forests cleared, atmospheric CO_2 content increases and the greenhouse effect intensifies. Forests are also capable of binding high amounts of dust (30–70 t/ha), thus contributing to cleaning the air from pollutants (Kerényi 2003).

Wood is one of the civilisation's most important *raw materials* and energy sources. Finally, forests play an important role in *recreation* as well. Their favourable climatic impacts advance the regeneration of the human body as well as satisfy the soul.

About 2,000–3,000 years ago, along with the development of transportation (shipbuilding, the appearance of wheeled carts), the destruction of temperate forests started to take place in the Mediterranean. (Many researchers associate the intensive transgression of almost all river deltas into the Mediterranean Sea and the silting of a number of antique harbours (Ephesos, Miletos, Ravenna, etc.) with the increased amount of debris resulting from deforestation.) Widespread grazing also impeded the regeneration of forests. Roughly 200 years ago, the majority of the forests in most of Europe and in Southeast Asia underwent the same process, partly due to the population growth and the demands by the developing industry. The natural forest cover fell victim to European settlers gaining ground in North America ca. 100 years ago (Plate 8.1). By the 20th century, deforestation became focused on the tropics. Today the destruction of the tropical rain forests takes place at an intensity never witnessed before, for which the primary explanation is given by the economic and social backwardness and the financial exposedness of the tropical countries.

Ten thousand years ago, an area of approximately 62.2 million square kilometres was covered by natural forests. (According to the definition by FAO, forests are areas at least 0.5 ha in size and in which at least 10% of the area is covered by foliage.) The area of natural forests, which represented nearly 42% of the land surface, dropped to 38.7 million square kilometres (26%) (Rakonczai 2003; Fig. 8.1). Each year, ca. 3 billion tonnes of wood is used. More wood is used in Europe for furniture, construction material, firewood and paper than metals for any other purposes.

The rate of deforestation at the beginning of the 21st century is estimated to be 16 million hectares per year. Of this, about 14 million hectares are cleared in the tropics. As a result of afforestation, the net loss of forested areas is 'only' ca. 10 million hectares.

Plate 8.1 A typical deforested landscape in the Midlands of England (Karancsi 1998)

The uppermost map in Fig. 8.1 indicates former natural forested areas, the middle one the total area of present-day natural and planted forests whereas the lower one indicates the extension of natural forests now. Monoculture, i.e. when only a single tree species is planted over large areas, is a feature of artificially planted 'cultivated forests'. With the trees of such forests becoming exploitable at the same time, clear-cutting, entailing intensified erosion processes (Plate 8.2) and land degradation (Plate 8.3) seem to be nearly unavoidable. In times of forest renewals, for economic reasons, alien species are often planted replacing indigenous ones with mostly negative impacts on the whole ecosystem (soil acidification, severe species degradation, etc.).

Hereafter, an overview will be provided on the relevance of *tropical rainforests* and on the consequences of clear-cutting on their environment (Fig. 8.2).

These forests with complicated vertical structures are extremely abundant in species. In certain regions of Amazonia, a single hectare contains as many plant and animal species as the total of European forests. (Here, as many as 2,000 tree species can be present on 1 ha in contrast to the forests of the moderate climate where only a maximum of 20 tree species can be found.) This is the richest ecosystem of our planet. At its greatest extension, it is found in the Amazon and Congo basins and also in Southeast Asia and Central America.

Tropical rainforests are also cleared to provide area for crop cultivation and animal husbandry as well as for the trading of tropical wood.

During forest clearance, first the multi-storied abundant foliage, responsible for the transpiration of large amounts of water, is removed. As the transpiration surface is reduced, the moisture content of the air drops and this results in lesser and more sporadic rainfall. The existence of a tropical forest is based on rainfall. With decreasing amounts of rainfall, the vegetation also begins to grow less dense and

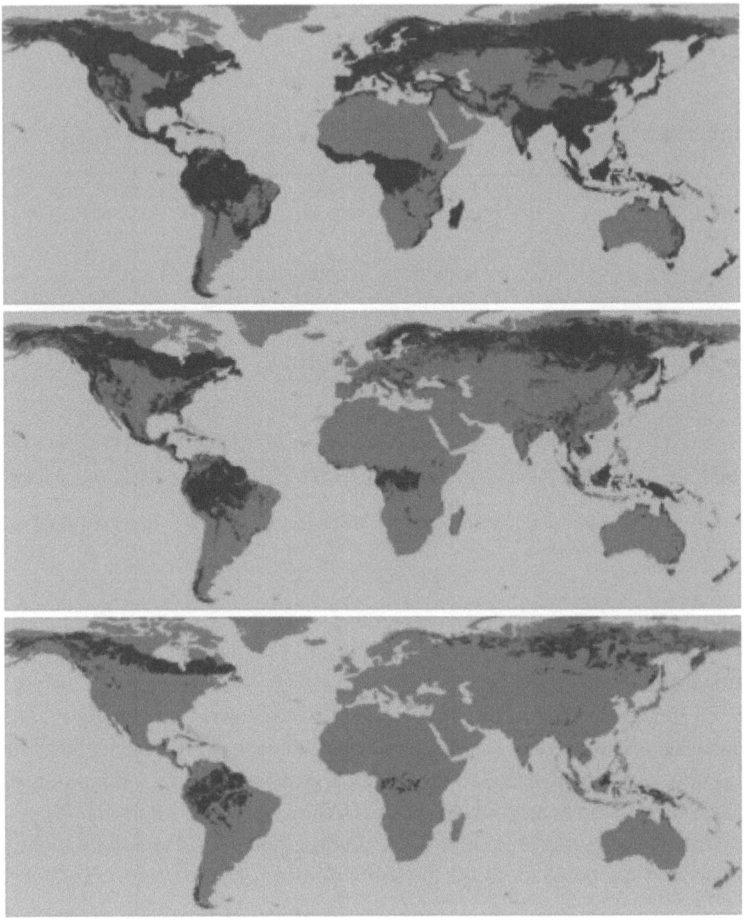

Fig. 8.1 Forested areas in the past and at present (Bryant et al.1997)

degraded. With a sparse foliage, irradiation increases and, due to the more intensive vertical turbulent air movement, violent hailstorms become common.

The *water-retaining capacity* of the vegetation cover reduced by the clearance of rainforests will have serious consequences. The foliage, coats of moss and lichen as well as epiphyte plants of the rainforests intercept rainwater conveying it towards the soil slowly (by drops). On the contrary, rainwater in deforested areas reaches the ground surface unhindered, washing away soil particles (erosion). The otherwise thin soil of the rainforests is gullied by erosion into *badlands*. In deforested areas, the increasingly scarce vegetation will be composed of less demanding plant species. Violent hails, after a while, can entirely wash away the laterite soils from the bedrock and make revegetation impossible.

Plate 8.2 Gully development after clear-cutting at the western rim of the Medves Plateau (Karancsi 2003)

Plate 8.3 Typical 'badland' formation after deforestation at Kazár, North-Hungary (Karancsi 2006)

Intense rainfall will raise the water-level of rivers causing *floods*, which destroy agricultural lands and settlements along the river banks. Fluvial sediments will fill up lakes furthering large-scale eutrophication.

Tropical rainforests evenly use solar energy and reflect only 8–14%. The albedo of the bushland formed after forest clearance is 15–20%, whereas that of an eroded surface can exceed 30%. Intensive inward and outward radiation also mean a great

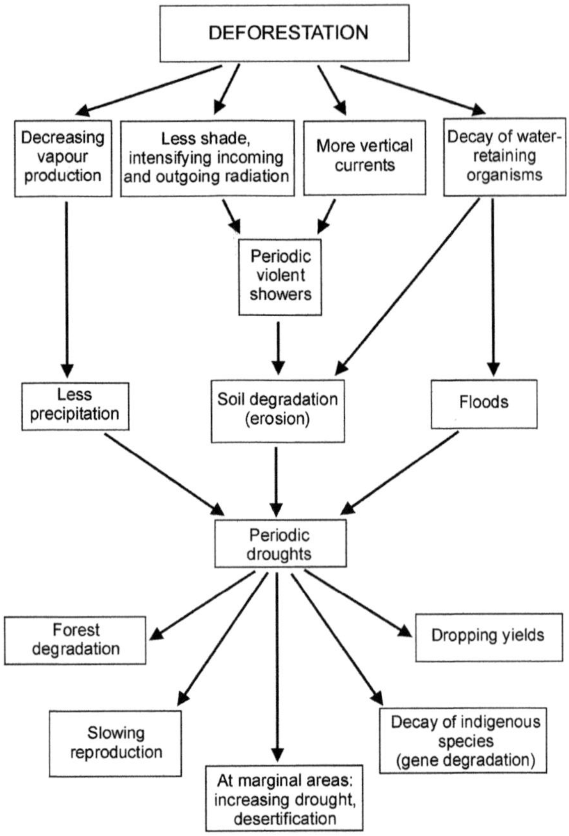

Fig. 8.2 The impact of tropical forest clearance on the environment (Balázs 1990)

fluctuation of temperature that is not tolerated by plants accustomed to evenly warm and humid climate. Further species will be extinct.

The clearance of tropical rainforests also has an influence over further regions, where the amount of precipitation is also reduced and seasonal droughts become more frequent there. Thus former deciduous closed forests are replaced by savannas with groves, wooded savannas become treeless grasslands, whereas in former grasslands, grass becomes sparse and desertification begins to take place. This process has a disastrous impact on farming. Farmers are forced to move closer to areas under equatorial climate and with a higher amount of precipitation, where, on the contrary, endowments for production (shallow soils poor in nutrients) are less favourable and new agricultural areas are gained by forest clearance. Stock keepers are also forced to leave their pastures abandoned under desertification and have to form new grazing lands for either farmers or forested areas.

Tropical rainforest clearance is also associated with the destruction of local peoples' lifestyles.

8.2 History of Woodlands in the Carpathian Basin

The Holocene was the period when the Carpathian Basin became forested and the stock-breeders and farmers appeared. This is, therefore, also the advent of deforestation. The parallel processes of the natural expansion of forests and anthropogenic impacts influenced the history of Hungary's forests in the past ca. 8,000 years. (Traces of the first settlements of farmers in Hungary are from the 6th millennium BC. The age of the earliest Neolithic site (Szeged-Gyálarét) is dated by the radiocarbon method at 7090 $\pm$ 100 BP (Bácskai 1982)). It is estimated that without human influence on the environment, the natural vegetation of Hungary at present would be forest steppe and more than 60% forested (Medzihradszky 1996).

Forested areas reached their greatest extension during the Paleolithic and the Iron Age, as climate provided the most favourable conditions to the expansion of woods while nature transformation by humankind were only of a minor scale at that time. During the Neolithic Period with more arid climate, there was a decrease in wooded areas, while grasslands expanded. Prior to the beginning of human landscape transformation, approximately 60% of Hungary was covered by forests. By the time of the Magyar Conquest, forest clearance reduced the cover to 43%. The country witnessed the lowest rate of forested areas (11.8%) in 1913 (the data is for the area of present-day Hungary). Today, forests cover more than 20% of the area; however, only one-third of this resembles former (potentially) natural forests whereas the rest is plantation or intensively transformed forests (Németh 1998). [Here, the notion of (potential) natural forest refers to the presence of zonal associations (Turkey oak-sessile oak, hornbeam-oak on higher terrains, and beech on cooler slopes with northern exposure) less disturbed by humans.]

According to the results of pollen analyses, the Hungarian low mountains could have been mostly covered by beech woods in the Subboreal (2,500–5,000 years ago) (Pócs 1981). In the Subatlantic, lasting to the present, such beech woods retreated to higher regions. Human appearance in the Carpathian Basin is estimated at about 6000 BC. Anthropogenic impacts were, in the Neolithic, restricted to rather small patches, landscape transformation only accelerated later. Farming and animal husbandry have been gaining ground at the expense of forests along the lowland margins from the Copper Age. Forest clearance involved accelerated soil erosion and renewed movements of wind-blown sand.

8.3 Case Study: Geomorphological Impacts of Deforestation Through the Example of the Medves Region

The research area of 32 km^2, henceforth called the Medves Region, is part of one of the micro-region of the North Hungarian Mountains, called the Medves Plateau, Central Europe's largest basalt plateau (with an area of 13 km^2, of which an 8 km^2 section falls within the territory of Hungary) and the adjacent regions with a more varied morphology and basalt peaks of higher elevations (e.g. Salgó – 625 m,

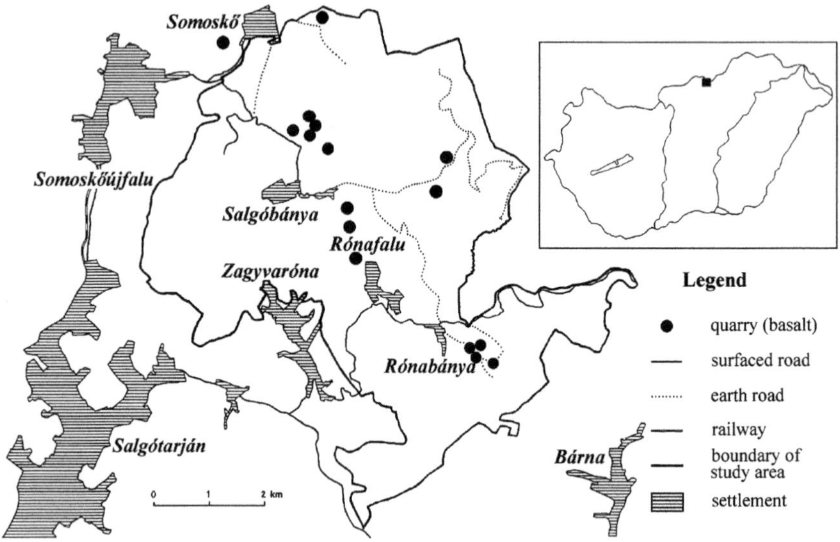

Fig. 8.3 The Medves Region with the study area (Karancsi 2005)

Szilváskő – 628 m) (Horváth et al. 1997; Karancsi 1998a,b). This area with diversified morphology neighbouring Slovakia has been part of the Karancs–Medves Landscape Protection area since 1989 (Fig. 8.3).

8.3.1 A Historical Review of Agricultural Landscape Alterations

To understand anthropogenic impacts, the temporal and spatial variations and human activities in the region have to be studied. In lack of a comprehensive archaeological survey, general data as well as written and mapped sources found in archives provide information on the area (Erdősi 1978; Gazdag 1964).

The majority of the research area has been used by agriculture and forestry for a long time (Dornyay 1928). In the Copper Age, beginning here around 2500 BC, an ethnic group similar to the Badenian Culture populated the area. Unlike Stone Age people, they were primarily stock keepers, thus were often forced to change their locations. During herding, new pastures were created by, in addition to migration, forest clearance, i.e. the appearance of anthropogenic impacts can be estimated by and large to this period in the region. Migratory herding caused trampling; deforestation accelerated erosion, in other words, gradual land degradation and dissection.

The Bronze Age (1800–750 BC) witnessed population growth. Among the people of the Iron Age starting around 750 BC, Scythians involved in animal husbandry (mainly of horses dominated the area for nearly 300 years).

This group carried out clearance farming around its settlements of shorter or longer duration. They were followed by the *Celts* (250 BC) who introduced farming by applying animal yoke-power (by using a wooden plough with iron shoe), changing the soil structure of cultivated areas even to a depth of 20–30 cm. This period is thus considered to be the first period of remarkable *environmental transformation* for agricultural purposes which led to irreversible changes, e.g. soil erosion (Karancsi 1999; Karancsi and Mucsi 1999).

Following the Magyar Conquest, masses of Hungarians continued to settle in the early 11th century while the population previously mainly involved in herding turned to farming (Dornyay 1948). With the emerging kingdom, of the ancient common properties, first royal forest manors or forest land-stewardships got separated. They were the most relevant estates.

From *14th and 15th century* sources, one can conclude that the region was economically prosperous. An increasing amount of land was continuously under cultivation; this led to the disappearance of natural (mainly forest) vegetation and *accelerated erosion* processes (Gunst 1970).

At the time of the *Ottoman* occupation, a relatively large amount of land was available for *extensive animal husbandry*. This type of land use with a area demand greater than intensive husbandry required further forest clearance.

Following the defeat of the Rákóczi War of Independence (1703–1711), most of the country was fallow lands or forests. The *expansion of arable land*, however, grew by the conversion of grazing lands and further deforestation during the *18th century*. Peasants, by hard work, recaptured abandoned arable lands as well as opened up new ones by deforestation (e.g. in the Somoskő Estate, half of the arable land was cleared land in 1717; also in the 1700s, significant forest clearance took place at the outskirts of Rónafalu in the southern part of the Medves Plateau.) A section of the area gained was thought to be seeded by row crops, later by cereals, while the rest was used by grazing.

As a consequence of the large-scale deforestation, the amount of water running off from the catchment areas of mountain regions increased. Erosion and floods became more common in the lowlands.

Runoff regulation by forests also has its limitation. For old stands and absolutely dry soils, retained runoff is estimated to be 175 mm per day. When rainfall triggering flood waves occurs following humid weather, this value can drop to as low as 100 mm. The most unfavourable *river regime* is present along watercourses of valleys with catchment areas affected by clear-cutting. In such areas, flood discharge can increase even by 60–140%, while maximum discharge by 200–400%. It was observed that the amount of runoff, with identical slope conditions is 1.5 times higher when pine forests are clear-cut than when beech woods are cleared. (Runoff variation depends on the thickness of forest undergrowth (litter) as well as on soil quality and depth.) For selective logging, the rate of this unfavourable impact can drop to as low as its one-tenth (Illés and Konecsny 2000). Although the presence of forests reduces flood risk, it cannot be considered, by any means, as the primary or decisive means for flood control (Vágás 2001).

In 1729, the landowner György Szluha intended to impede damage caused by erosion in the surroundings of Salgótarján. Based on his field experience he ordered that deep furrows should be ploughed at the upper and lower ends of plough-lands, so that 'abrupt rainfall should not cause outwash and any damage in lands' (Szabó 1972).

Meadows adjacent to Salgótarján were located to the north and south from the then village in the valley, in a width of 200–300 m along the stream. Due to the unreasonable deforestation carried out, because of the frequent floods resulting from spring thaw and hails, *soil* was removed from steep, bare hillsides and *accumulated in the valleys*. As read in contemporary description, lime and willow trees were almost buried in mud in the valley of the Tarján stream (Fig. 8.4).

In the village of Salgótarján, the greatest damage was caused by the floods of 1840, 1854, 1870, 1873 and 1926. The disastrous flood following the downpour on 25th May 1926 proved the necessity of carrying out channelisation, which along with stream regulation, has gained importance in urban development plans since the 1930s. This reduced flood hazard in Salgótarján. Meadows have become silted, continuously degraded, which was the primary reason to dropping animal-stocks. Arable lands were in most cases adjacent to meadows on hillsides and water gaps around the village (Ponyi and Diós valleys, Baglyasalja) while, pastures of almost bare surface were found on the mounds above. This distribution is the most unfavourable as it is exposed to large-scale erosion hazard.

In the mid-18th century, Ferenc Szluha landowner prohibited oak-masting and wood-cutting, allowed in manor forests, resulting in a drop in pig-keeping. Then sheep-keeping became widespread and forests in the surroundings of settlements were almost entirely cleared.

There were intentions to develop more intensive branches of agriculture as orchards in the 19th century. Consequently, although there was an increase in the fruit-tree stands, fruit production has never become really significant.

The structure of agriculture, following the mid-1920s has witnessed only small-scale changes. In addition to diminishing fallow land, however still extensive, bare eroded areas resulting from unplanned deforestation further increased.

By the establishment of nature conservation areas (since 1964), land-use regulations have become stricter. Due to the unfavourable soil conditions (mostly shallow, brown forest soils with clay illuviation), cooperatives produced fodder-crops for their animal stocks in addition to the profitable forestry since the 1970s.

Today, some arable lands are replaced by planted grasslands under the management of the Bükk National Park Directorate (BNPD). They were planted more than 10 years ago, and spontaneous afforestation has started on them. Such areas are found around Somoskőújfalu (Judik 2000).

8.3.2 Changes in the Forested Areas in the Medves Region

Humans do not only utilise timber and other forest products but also exploit forest areas. Forests were regarded as cheap reserve areas for urban sprawl, expansion of

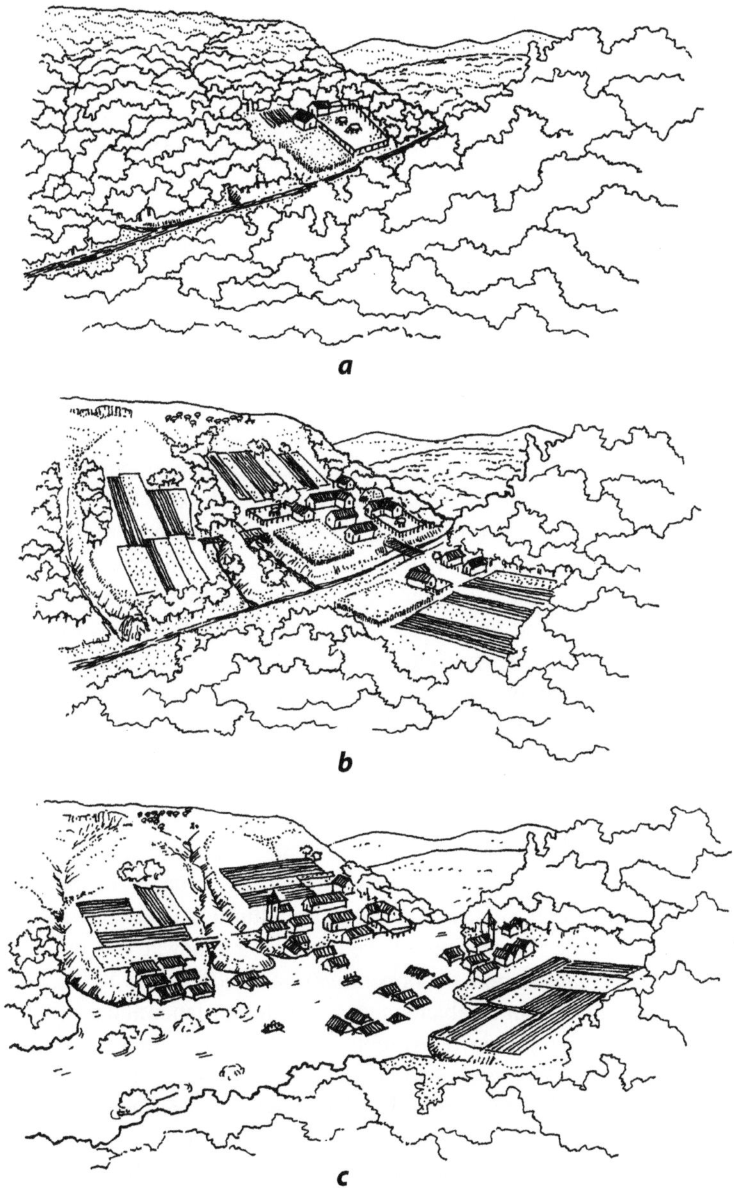

Fig. 8.4 Human impacts on the physical environment (Karancsi and Mucsi 1999). a. Human settlement; b. Expansion of agricultural land; c. Flooding

arable land, pastures or mines. By depriving the area from its primary vegetation cover, irreversible processes often started.

In lack of bogs suitable for pollen analysis, the question what the *potential* (primary) *vegetation* in the Medves Region prior to intensive deforestation was, is hard to answer (Zólyomi 1981; Csiky 1997). The only buried charcoal sample found is, according to absolute dating, maximum 200 years old.

According to archive sources, in the Middle Ages the Medves Plateau, now mostly deforested, was covered by immense forests whose exact species composition is not known. Documents from the 14th and 15th centuries mention that 'the Medves is covered by extensive beech wood' while a manuscript from the early 18th century – according to which the surroundings of the Medves (in contrast to the Cserhát Hills) were characterised by oak and extensive beech woods – provide a more detailed picture on the region's forests (Csiky et al. 2001).

Although, by the early 19th century, landowners started to realise the necessity of professional forest management, they seemed to have lost the opportunity as by the middle of the century, the extension of forests proper was rather limited and their quality poor around Salgótarján.

Regulations on forest conservation and the maintenance of stands are included in forestry plans for more than 200 years in Hungary. The concept of *shelter-wood* referring to forest sections prohibited to cut is already present in the documents. They regarded to be beneficial as their roots hinder soil erosion; they protect mountain roads and other structures. Shelter-woods around buildings or agricultural areas also have a windbreak and temperature-adjusting effect. Forests emitting oxygen are capable of absorbing a large amount of pollutants (estimated to be 30–70 t of dust and air-pollutants by each hectare of forests). Therefore the role of forests around industrial sites, waste heaps and noisy, air-polluting factories is remarkable (Mészöly 1981).

Planned forest management in the Medves Region only began in the late 19th century. This regeneration involved mainly Turkey oak-sessile oak-woods; later the reforestation of areas with degraded soils was also launched mostly by the then favoured lime (as both species excellently fulfilled all demands of that time: providing sufficient amounts of firewood and agricultural timber). Under the trees, grass perfectly suitable for grazing evolved and their renewal was relatively simple. Following the establishment of cooperatives, in the 1960s and 1970s, large-scale afforestation was carried out on abandoned, steep plough-lands and pastures, mostly on degraded forest soils, primarily by Scots pine *(Pinus sylvestris)*.

As a result of the plantations carried out then and since, by now the original (potential) vegetations has been entirely exchanged, as more than one-third of the forest area are plantations established over the past 40 years.

The most important indigenous species for forestry in the Medves Region are beech *(Fagus sylvatica)*, sessile oak *(Quercus petraea)*, Turkey oak *(Quercus cerris)* and hornbeam *(Carpinus betulus)* (Zólyomi 1981,1995). On the warmer hillsides of southern exposure of the study area, Turkey oak-woods stands are found whereas sessile oak-woods predominate at locations with northern exposure. On cooler slopes above 500 m as well as in deep-cut valleys, beech woods without undergrowth and with deep litter are formed (Fancsik 1989).

There are alder forests in the valleys. *Robinia pseudoacacia*, planted around the turn of the 19th and 20th centuries mainly to replace diminished oak-woods, is apparent nearly everywhere. Now, at the site of the manifold cut stands, bushes are found.

Relatively large areas are covered by also planted by Scots pine and Austrian pine (*Pinus nigra*) patches jointing Turkey oak-sessile oak forests (*Quercetum petraeae-cerris*) appear up to the elevation of 400–500 m. They were planted on mainly eroded soils with shallow topsoil. Young pines are often found in new plantations, regarded as pioneer forests. On eroded lands and uncultivated low-quality dry sand soils, mainly Scots pine and Austrian pine can be planted. During one rotation period (50–60 years), soil nutrient content increases and becomes suitable for planting more exacting and valuable stands. The uniformity of plateau margin forests is disrupted by patch-like wedging arable fields which are linked by narrow forest belts (as ecological corridors).

Comparing the available topographic maps, the changes in forested areas are reconstructed (Fig. 8.5).

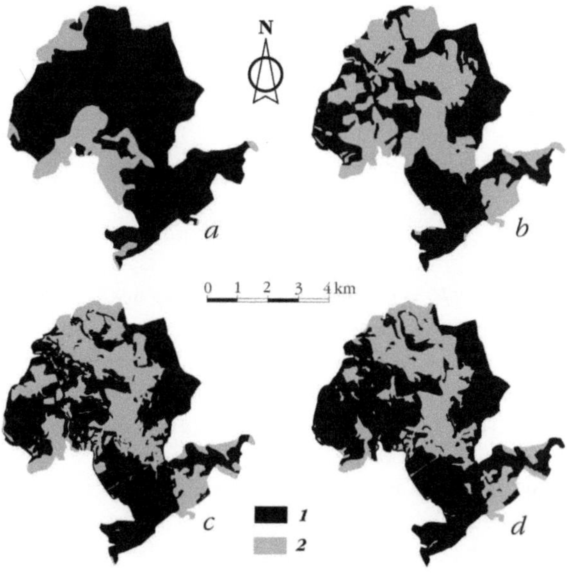

Fig. 8.5 Changes in the forested areas in the Medves Region (according to the data of (**a**) the 1st military survey in 1782, (**b**) the soil map by Kreybig, L. in 1922, (**c**) the topographic map at the scale of 1:10,000 from 1966 and (**d**) the military map at the scale of 1:25,000 from 1988)

The most detailed maps of the forests in the region are obviously supplied by forestry mapping repeated about every 10 years. Having this compared to the forest patches on the map from 1988, it can be concluded that 85% of the area in the past 10 years has hardly changed. 58% of the study area is still covered by forests whereas 27% is mostly used as hay meadows. In total, the rate of clearances exceeds that of new plantations by 2%. It is favourable that, despite having less new forest patches on the map, their extension is significantly larger than the mostly dispersed small clearances.

In the evolution of most of the semi-natural associations, anthropogenic activity played an important role. Many of the present-day forests of good condition have grown at the sites of pastures, while grasslands that have become quasi-natural by today, evolved at the sites of former forest clearances. Although, by human contribution, the original forests have been replaced in almost the entire area of the Medves Region, at 62% of the total forested areas, zonal associations are found. Oak-woods have the greatest share of area (702.85 ha, 34%) (Table 8.1). The oak stands include Turkey oak-sessile oak *(Q. petraeae-cerris)*, hornbeam-oak *(Querco petraeae-Carpinetum)*, red oak *(Quercus rubra)*, pedunculate oak *(Quercus robur)*, sessile oak *(Q. petraea)* and pubescent oak *(Quercus pubescens)* (Csiky 1997).

Table 8.1 Tree species distribution in the forests of the Medves Region

Wood species	Area (ha)	Area (%)	Forest patch
Oak	702.85	34.00	113
Beech	557.62	27.00	81
Pine	328.33	16.00	83
Mixed	249.01	12.00	54
Pseudoacacia	205.94	10.00	24
Alder	8.14	0.40	4
Trembling aspen	7.19	0.35	1
Birch	3.67	0.17	2
Goat-willow	1.30	0.05	1
Ash	1.10	0.03	1
Total	2,065.15	100.00	364

Beech *(F. sylvatica)* stands are the next widespread (557.62 ha, 27%). Such forest sections are characterised by less diverse species composition (mixed with hornbeam). 16% (328.33 ha) of the total forest area is occupied by planted pine-woods. The pine stands include Douglas fir *(Pseudotsuga menziesii)*, Scots pine *(P. sylvestris)*, Austrian pine *(P. nigra)*, Norway spruce *(Picea excelsa)*, Eastern white pine *(Pinus strobus)* and a further 12% (249.01 ha) mixed forests, which include hedge maple *(Acer campestre)*, Norway maple *(Acer platanoides)* and small-leaved lime *(Tilia cordata)*. The penetration by *R. pseudoacacia* of 10% (205.94 ha), however, is considered to be unfavourable. Other wood species in the Medves Region, considering their area, are irrelevant as none of them approximates 1%. These are trembling aspens *(Populus tremula)*, silver birch *(Betula pendula)*, goat-willow *(Salix caprea)* and white willow *(Salix alba)* common at clearings and in new forests along the alders *(Almus glutinosa)* of creek banks and waterlogged areas. Ash *(Fraxinus excelsior)* also prefers more humid habitats.

Among the intensive landscape transforming human activities, *quarrying* should also be mentioned. Fortunately, the few decades since the quarries were closed have proved to be sufficient to the development of a variety of semi-natural vegetation at the areas impacted.

When studying the distribution of climazonal associations characteristic for the region as well as that of anthropogenic alien ones, the following conclusions can be made (Csiky et al. 2000): As most of the Turkey oak-sessile oak forests were located near the former settlements, attractive and old oak stands are relatively rare there. Moreover, most of the arable land, pastures and pseudoacacia woods are found replacing them as well as pine species planted later. Pseudoacacia and especially Scots pine and Austrian pine subsist in areas affected by erosion, with shallow topsoil and poor in nutrients. While soils are further degraded under pseudoacacia, ever more difficult to replace, under pines soil nutrient content improves in approximately 50 years, thus the area becomes suitable for planting more valuable tree species.

Rainwater is gathered (and remains for a shorter or longer period of time) in subsidences of the area formed by indirect anthropogenic impact, mostly by mining. This acidic stagnant water provides favourable living conditions for birches. The *birch wood circles* typical of the plateau were formed by human activities.

Turkey oak-sessile oak forests, formerly characteristic for the whole region, are still typical along the steep Eastern side of the plateau and its southern, less disturbed regions. The proximity to the Slovakian border also influenced the persistence of forests in the eastern side of the plateau. In areas of the highest elevation, above 500 m, beech becomes predominant. At the summit regions with gentle slopes (e.g. Szilváskő, Medves hilltop) submountain beech woods are typical, thus they are considered to be the uppermost zonal forest association in the Medves Region. Beech woods occur in the deep valleys of the hillsides with southern exposure already at the elevation of 300 m, while at northern hillsides as low as 200 m (Csiky et al. 2001).

Forests jointing the uniformity of arable lands and hay meadows have been sustained in the plateau, usually in areas of dissected surface with unfavourable agricultural potential.

8.3.3 Vegetation Related Environmental Conflicts

The Medves Region witnessed significant deforestations in the past centuries. Cultivated areas rapidly became degraded and were abandoned and later used as pastures. Due to the lack of forest cover and as a result of erosion, *deep gullies* formed in the loose soils. Flatter valley sections were filled up by alluvium and waterlogged meadows resulted. Overgrazing led to soil degradation both on slopes (most of which were later forested) and in forests. Oak-masting, however, made the renewal of more valuable wood species more complicated or even impossible. This peculiarly dysfunctional, multiple 'use' of forests inevitably caused a decrease in the ecological and (as seen today) economic values of forests.

A typical data for damage caused by agriculture, upsetting the ecological balance, can be cited. The use of chemicals in the 1970s, both gophers and the saker *(Falco cherrug)* preying on gophers have disappeared from the plateau.

The pastures of the plateau are practically the most suitable areas for sheep-breeding (however, in the late 1990s, cattle herds were also pastured) only in summer. However *grazing* also had environmental consequences. Earlier it was typical that the area was fenced around and grazed until it became entirely barren: Then a new area was selected for grazing. Fortunately, this type of intensive pasturing does not occur in the region any more.

Imbalance is also present in water resources due to deforestation. Rainwater rapidly ran off and springs have run dry; watercourses have become desiccated and groundwater recharge has also become irregular.

Forest clearance and pasturing taking place for centuries caused the spatial expansion of the grasslands under Turkey oak-sessile oak-woods, pubescent oak forests and sandstone cliffs to regions formerly occupied by other associations. This unfavourable alteration is still taking place today in the wake of aridification and soil degradation.

Former and present forestry seriously impacts the habitats of the study area. In many areas, forests composed of alien species, which have replaced native trees. Not only the spatial expansion of pines is unfavourable but the pseudoacacia also cause problems. The most significant change in the distribution of tree species lately regards Scots pine, the species mostly used for plantations. A minor growth has also been caused by stand transformations. The growth in the area of Norway spruce exclusively resulted from the alteration of degraded forests. The decrease of Turkey oak stands is favourable as it is replaced by more valuable species. There has been practically no change in the area of the beech forests. However, the spatial decrease of oak is unfavourable.

Light conditions and nitrogen abundance of pseudoacacia stands as the zones of spreading of nitrophile weeds also cause the degradation of the surrounding areas. By its aggressive dispersion, *pseudoacacia invades* nearly all types of vegetation and when proliferated will bring forth their intensive degradation (Harmos 2000). The area and natural dispersion of pseudoacacia increased because of the plantations, despite the stand alteration carried out to impair both pseudoacacia and Norway spruce. Most of them are usually bushlands of reduced timber productivity. They have to be replaced sooner or later. This is especially the case for pseudoacacia woods on heavy soils and in frost hollows but in good habitats. They are extremely diverse in age, furthered by forestations carried out in the past 40 years as well as by species with various rotation periods.

Forests also provide shelter to the primary fauna as well. Recently, there has been a slight increase in the numbers of *game*, therefore the *damage* caused by them (e.g. trampling and chewing) has also become more apparent, even resulting in the disappearance of more susceptible species from both forests and grasslands of certain regions. Degradation is also caused by treading and littering at frequently visited areas, especially at extremely susceptible grasslands.

References

Bácskay E (1982) A magyarországi holocén sztratigráfia régészeti dokumentációs pontjai a Dunántúlon. Korai és középső neolítikum. (Archaeological documentation points of the Hungarian Holocene stratigraphy in the Trans-Danubian region. Early and middle Neolithic) M. Áll Földt Int Évi Jel 1980-ból, 543–552

Balázs D (1990) Az őserdők világa (The World of Rainforests). Móra Ferenc Könyvkiadó, Budapest

Bryant D, Nielsen D, Tangley L (1997) The Last Frontier Forests: Ecosystems and Economies on the Edge – WRI, Washington, DC

Csiky J (1997) A Medves növényzete (Vegetation in the Medves). In: Horváth G, Munkácsy B, Pintér Z, Csiky J, Karancsi Z, Prakfalvi P. A Medves Földr Ért 46 (3–4): 242–244

Csiky J, Bódis K, Karancsi Z (2000) GIS módszerek alkalmazása a zonális vegetáció kimutatására egy erdőtlen fennsík példáján (GIS applications to detect zonal vegetation by the example of a deforested plateau). ACTA Biologica Debrecina Oecol Hung 11 (1): 208

Csiky J, Drexler Sz, Fancsik J, Horváth G, Karancsi Z (2001) Turizmus és természetvédelem kapcsolata a Medves-fennsíkon és közvetlen környékén (Relationship of Tourism and Nature Conservation on the Medves Plateau and Its Very Surroundings). (Medvesi tanulmányok I.) A Magyar Madártani és Természetvédelmi Egyesület Nógrádi Csoportjának tanulmánya. Kézirat, Salgótarján

Dornyay B (1928) Salgótarján történetéhez (To the History of Salgótarján). A Salgótarjáni Chorin Ferenc Reálgimnázium Értesítője az 1927/28. tanévre. 3–19

Dornyay B (1948) Salgótarján és vidéke a Honfoglalás korában (Salgótarján and its surroundings at the time of the Conquest of Hungary). Magánkiadás Salgótarján

Erdősi F (1978) Történelmi források és térképek szerepe a környezetben antropogén hatásra végbement változások földrajzi vizsgálatakor. (The role of historical sources and maps in the geographical study of anthropogenic changes). Földr Közl 102 (2): 118–127

Fancsik J (ed.) (1989) Nógrád Megye védett természeti értékei (Protected Natural Values in the County of Nógrád). Nógrád Megyei Tanács kiadása, Salgótarján

Gajzágó A (ed.) (1962) A salgótarjáni iparvidék (The Industrial Region of Salgótarján). Nógrád County Workers Movement Museum, Salgótarján

Gazdag L (1964) Nógrád megye térképei (Maps of Nógrád County). Nógrád megyei Múzeumi Füzetek 6, Nógrád Megyei Múzeumok, Balassagyarmat

Gunst P (1970) A mezőgazdasági termelés története Magyarországon (The History of Agriculture in Hungary). Akadémiai kiadó, Budapest

Harmos K (2000) Társulások természetességi megoszlása a Karancs-Meves Tájvédelmi Körzet területén (Natality Distribution of Associations in the Karancs-Medves Landscape Protection Area). In: A Karancs-Medves Tájvédelmi Körzet kezelési terve 2000–2010

Horváth G, Munkácsy B, Pintér Z, Csiky J, Karancsi Z, Prakfalvi P (1997). A Medves (The Medves plateau). Földr Ért 46 (3–4): 217–248

Illés L, Konecsny K (2000) Az erdő hidrológiai hatása az árvizek kialakulására a felső-Tisza vízgyűjtőben (Hydrologic impact of forests on flood formation in the catchment area of the upper Tisza). Vízügyi Közl 82 (2): 167–197

Judik B (2000) Karancs-Medves Tájvédelmi Körzet kezelési terve 2000–2010 (Management Plan of the Karancs-Medves Landscape Protection Area between 2000 and 2010). Manuscript, Salgótarján

Karancsi Z (1998a) Történeti áttekintés Salgótarján térségének antropogén eredetű környezetváltozásáról (A historical review on the anthropogenic environmental changes in the Salgótarján region). Geográfus doktoranduszok I. országos konferenciájának kiadványában. In: Tóth J, Wilhelm Z (eds.) (1996) A társadalmi-gazdasági aktivitás területi-környezeti problémái. JPTE, Pécs

Karancsi Z (1998b) Az ember környezet-átalakító tevékenysége a Karancs-Medves Tájvédelmi Körzet területén (Anthropogenic Environment-Transformation Actions in the Karancs-Medves Landscape Protection Area). Geográfus doktoranduszok II. országos konferenciájának kiadványában, Budapest (1997. November)

Karancsi Z (1999) A mezőgazdasági táj változása a Medves-térségben (Changes of the Agricultural Landscape in the Medves Region). A táj változásai a Kárpát-medencében. A II. tájtörténeti konferencia (Nyíregyháza, 4–6th November 1998) kiadványa

Karancsi Z, Mucsi L (1999) Az emberi tevékenység hatása a Medves-régió területén (Anthropogenic impacts in the area of the Medves region). Magy Tud 44 (9):1140–1147

Kaulfusz J (1854) Nógrádi erdők (Forests in the County of Nógrád). Gazd Lapok
Kerényi A (2003) Környezettan (Ecology). Mezőgazda Kiadó, Budapest
Medzihradszky Zs (1996) A magyarországi erdők rövid története (A short history of forests in Hungary). Földr Közl 120 (44) (2–3): 181–186
Mészöly Gy (ed.) (1981) Parkerdők Magyarországon (Park Forests in Hungary). Natura, Budapest
Németh F (1998) Magyarország erdőterületeinek változása 1100 év alatt (Changes in the forested areas of Hungary in 1100 years). Erdészeti Kutatások 88: 65–89
Pócs T (1981) A magyar flóra és növénytakaró kialakulása (Evolution of Hungary's flora and vegetation cover). In: Hortobágyi T, Simon T (eds.), Növényföldrajz, társulástan és ökológia. Tankönyvkiadó, Budapest
Rakonczai J (2003) Globális környezeti problémák (Global Environmental Problems). Lazi Könyvkiadó, Szeged
Szabó B (ed.) (1972) Salgótarján története (The History of Salgótarján) – Salgótarjáni Városi Tanács V.B., Salgótarján
Vágás I (2001) Az ezredforduló árhullámai a Tiszán (Floodwaves at the turn of the millennia on the river Tisza). Magy Tud 8: 958–966
Zólyomi B (1981) Magyarország természetes növénytakarója (The Primary Vegetation Cover of Hungary). Kartográfiai V., Budapest
Zólyomi B (1995) A Kárpát-medence és környékének potenciális növénytakarója az 1. évezred fordulója körül (Potential vegetation cover in the Carpathian Basin and its surroundings around the turning of the 1st millennium). História 17 (3): 4–5

Chapter 9
Quarrying and Other Minerals

Lóránt Dávid

Abstract This chapter is an introduction to the significance of quarrying from the point of view of anthropogenic geomorphology. As a consequence of the extraction of mineral raw materials, so-called 'mining landscapes' have emerged since the 19th century. The spatial distribution of quarrying of aggregates, and the characteristics and classification of the resulting features are described at macro-, meso- and microscales. Quarry walls and floors and debris aprons are distinguished at almost every extraction site. The morphological components of accumulated macroforms are plateaux and slopes (accumulated mesoforms). Common excavated microforms of quarrying are rock counterforts, rock benches, out-weathered quarry columns, pinnacles and pillars. Finally, international and Hungarian case studies illustrate some aspects of the opening and after-use of mining sites in order to observe how abandoned quarries can be turned into 'environmental values', and used as possible sites for exhibitions or for regional and tourism development projects.

Keywords Mining landscapes · Stone quarrying · Quarrying landforms · After-use

9.1 Introduction

There is no need to explain in detail *the close relationship between mining activities and geology/geomorphology*. However, it should be mentioned that researchers only became interested in the problems of geomorphic impact at a rather late stage of evolution of these sciences. It is well illustrated by Fig. 9.1 that research on landscape alterations caused by raw material extraction only dates back to the 1960s.

L. Dávid (✉)
Department of Tourism and Regional Development, Károly Róbert College, Mátrai út 36, 3200 Gyöngyös, Hungary
e-mail: davidlo@karolyrobert.hu

J. Szabó et al. (eds.), *Anthropogenic Geomorphology*,
DOI 10.1007/978-90-481-3058-0_9,

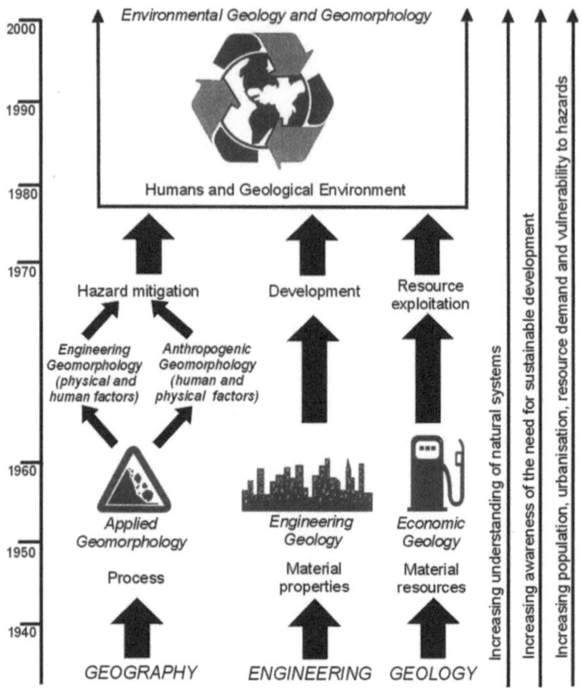

Fig. 9.1 The evolution and differentiation of earth sciences and their relation to environmental issues (after Bennett and Doyle 1999, modified by Dávid and Baros 2006)

Two million years ago, the activity now called mining was just occasional collection of rocks. This early extraction of gravels and stones was later replaced by conscious exploitation of minerals for flint tools, dyes and other purposes. According to the archaeological findings of the oldest flint quarries, this has at least a 100–120 thousand years' history in Hungary. A well-studied quarry of dyestuff at Lovas near Veszprém, as well as the flint quarry discovered in the area of the Tűzköves (Avas, Miskolc) are dated to be several tens of thousand years (40–110 thousand years) old (Simán 1995). Shelters of the primitive man were often carved out of loess. Flint quarries of the Neolithic Age are 6–8 thousand years old in Hungary (Sümeg-Mogyorós Hill, Tata-Kálváriadomb, Miskolc-Avashegy) (Simán 1995). Hollows of the oldest salt mines in the Pannonian Basin can be of similar age. In the 2nd century, the famous mines of the Transylvanian (Dacia) 'Gold Rectangle' (Abrudbánya, Nagyág, Zalatna and Verespatak) were obtained by the Romans and were cultivated through 170 years under imperial rule. The mining of native copper, native gold, tin and antimony gradually displaced the mass use of flint tools. Ore and salt mines, based on their size, were much larger than the former hollows and resulted in caverns and later underground shafts.

In the 18th–19th centuries, a new energy source, coal, appeared in history and its mining resulted in landforms larger by orders of magnitude. Parallel to this, the development of mining technologies meant another step in the quarrying of building materials as well. The use of gunpowder and steam machines revolutionised the extraction of minerals and led to the emergence of 'mining landscapes'. In the world, the most commonly excavated raw materials for construction include those for cement and lime industry, building and ornamental stones, sand and gravel, as well as clays for ceramics. An introduction into the anthropogenic geomorphological significance of stone quarrying is followed by case studies on major landscape transformation resulting from the mining of other minerals.

9.2 Stone Quarrying

It can be claimed that the general spatial *distribution* of quarrying is fairly even in a sense that, *geological conditions* provided, there are hardly any mountain settlements without a quarry of some scale opened in their surroundings during their history. When quarrying also aims to reach markets to a greater distance, market regulators (economically exploitable supplies, transportation expenditure and possibility, etc.) become more important, thus in some cases, quarrying can show a rather high spatial concentration. The level of socio-economic development has been decisive for the quantity and quality of the material flow between the user and its environment.

In addition to the geological conditions, the site selection of quarries is also controlled by the *topography* of the area. Longwall face quarrying prevails on mountainous or hilly terrains whereas in flat areas deep mining is applied. However, intermediate types also occur occasionally. Exceptionally, closed work is applied, too, as in the case of Fertőrákos (NW Hungary). As far as longwall face quarrying is concerned, it is the topography that is transformed to a visible extent; face walls of several hundred metres length and of some ten metres height may result, depending on the technology applied (Fig. 9.2).

In cases when the rock material to be exploited is found under a flat or sloping surface, a quarry sunk in the surface has to be established. Such quarries are sometimes created through the lowering of the quarry floor by longwall face quarrying. If cover strata are too thick, extraction takes place from underground shafts or cavities. Apart from this, the characteristics of the quarried (metamorphic, igneous or sedimentary) rocks as well as adherence to various safety regulations are of decisive relevance. All of them may also have an influence on the evolving features.

9.2.1 Characterising and Classifying the Landforms of Quarrying

As a result of quarrying, the landscape undergoes fundamental and visible changes (Table 9.1).

The range of landforms resulting from excavation is classified into three main groups (Dávid and Patrick 1998; Karancsi 2000; Dávid 2000):

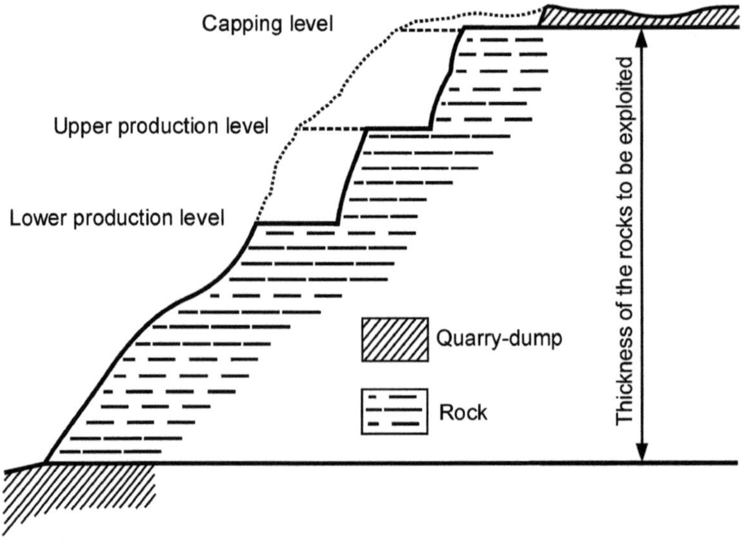

Fig. 9.2 Siting of a quarry of several production levels (Ozorai 1955)

- excavated ('negative') forms,
- accumulated ('positive') forms and
- forms destroyed by quarrying activities, which lead to the levelling of the surface, called planation in geography.

The geomorphologic study of quarrying features is undertaken in three categories, distinguished by origin and size (Fig. 9.3). It should be noted, however, that there are several other approaches to *classification* (Erdősi 1966, 1969, 1987; Karancsi 2000). One of them, for instance, is by quarry location relative to geological formations and surface macroforms (Erdősi 1987), whereas size categories taking the characteristics of the given area into account can also be set (Karancsi 2000).

Macroforms are the most obvious traces left behind by quarrying. Excavated macroforms may virtually be regarded as surfaces with material deficit (caverns) and are composed of smaller elements (excavated mesoforms). Quarry walls and floors and debris aprons are distinguished in almost every extraction site. The morphological components of accumulated macroforms are plateaux and slopes (accumulated mesoforms).

The surfaces of *mesoform* components can be divided into smaller and larger excavated depressions (possibly out-weathered sections) or accumulated elevations that are called microforms.

In addition to the influence of quarrying technology and working rate, the properties of features in all three categories are also controlled by the geological

Table 9.1 Landform-shaping role of quarrying activities (Dávid 2000)

Landform-Shaping Role of Quarrying Activities			
A. By the nature of the resulting surface features			
Excavated forms			Accumulated forms
Origin and size			
Excavated macroforms (surfaces with material deficit = caverns)			Accumulated macroforms (mine dumps) Cone-shaped Truncated cone-shaped Terraced
Quarrying technology			
Simple excavated type: excavation pit delph	Complex excavated type: horizon mining	Simple accumulated type: single quarry dump	Complex accumulated type: quarry dumps in groups
Excavated mesoforms Quarry wall Debris apron Quarry floor		Accumulated mesoforms Plateau Slope	
Microforms			
Excavated microforms: rock buttress and pillar pinnacles, rock benches, small shallow ponds	Microforms created as a result of natural processes: mass movements, gully erosion		Accumulated microforms: heap boulder
B. By the type of geotechnic activity			
Planation Abraiding		Filling up	

characteristics of the area (structure, bedding), the nature of the rocks and the natural processes affecting them.

9.2.2 Excavated (negative) Forms

The most common and simple type of excavated forms is an *excavation pit* or a cauldron in the surface. Excavated macroforms of quarrying origin usually appeared before accumulated forms, therefore examples of them can be found in the first period of quarrying history. Mainly in the form of small quarries they are found next to almost every town and village in mountainous areas.

The other type of excavated forms results from multi-levelled horizon mining (complex excavated type). It is increasingly typical in modern times. The technical condition for its occurrence was the increase in the capacity and efficiency of

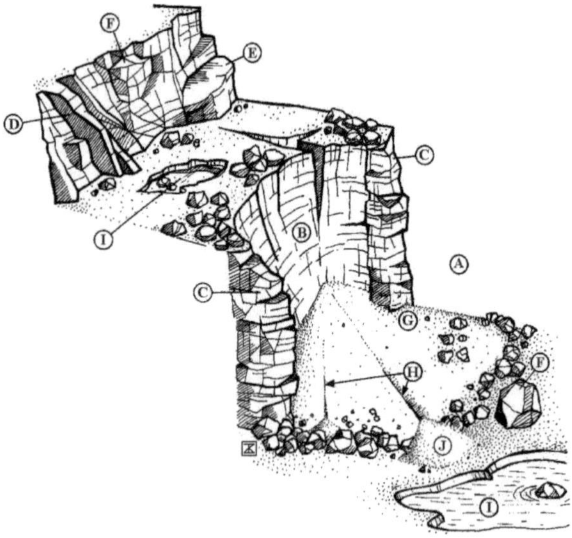

Fig. 9.3 Schematic layout of a stone quarry: (**a**) quarry floor, (**b**) quarry wall, (**c**) pillar, (**d**) rock buttress, (**e**) rock bench, (**f**) out-weathered rock, (**g**) talus slope, (**h**) rainwater groove, (**i**) depression with a small pond, (**j**) debris cone (Dávid and Karancsi 1999)

excavating equipment, while as far as geological conditions are concerned, it was favoured by the presence of thick strata.

Excavated mesoforms are composed of the following elements:

- Quarry wall: the steepest component, whose angle of inclination to the floor is determined by quarrying technology (blasting, hand or power excavation) as well as by rock quality. It is normally subvertical. The quarry floor is usually surrounded by walls on three sides.
- Debris cones, debris aprons: components with smaller angles of repose lying at the foot of quarry walls. Their materials partly derive from quarry working and partly from natural processes (rockfalls). They are initially developed by accumulation but their origin is linked to excavation activities. As material accumulates in debris cones, they may coalesce to form a continuous debris apron.
- Quarry floor: an approximately flat ground surface surrounded by walls and debris aprons, including a range of features (accumulations of quarry material, quarry heaps, pillars, etc.).

Common *excavated microforms* of quarrying are rock counterforts, rock benches, out-weathered quarry columns, pinnacles and pillars. The pillars are basically transitional features between excavated and accumulated forms as being the positive remnants of quarrying. They may resist the damaging effects of natural processes and talus slopes of various sizes are found in front of them. Precipitation water may collect in small shallow ponds in the depressions of the quarry floor.

9.2.3 Accumulated (Positive) Forms

Accumulated macroforms are called quarry dumps. They are formed through the accumulation of waste, which is currently of no value from an economic point of view (Fig. 9.4). During open-cast mining, dumps of various origin are heaped. By the removal of burden from above the material to be excavated, a significant amount of so-called sheathing dump is created. This material (interstage and plant dump) can also be a result of the extraction and processing of the material, i.e. during grinding or crushing. The granulometric composition of quarry dumps is rather diverse, being influenced not only by geological conditions but also by the method of processing. There can also be different shapes of dumps, as curve-, fan- and round-shaped dumps created at the end of bankfills are distinguished. In addition, temporary storage of the quarry material also has to be referred to this group. They are found isolated (simple accumulated type) or in groups (complex accumulated type).

The shape of a positive form is determined by several factors: the original ground surface, the mode of accumulation and the physical features of the dumped material. Cone-shaped, truncated cone-shaped and terraced dumps are the most common.

Common components of *accumulated mesoforms* are

- Plateau: the relatively flat ground surface surrounded by the slopes of dumps. Its extent is determined by the type of the dump. The largest plateaux can be found on terraced dumps, while the plateaux of truncated cone-shaped dumps are usually smaller.
- Slope: the sloping ground surface which surrounds the plateau or the peak in the case of a cone-shaped dump. Its inclination varies on a wide range depending on the mode of accumulation, the nature of the dumped material and the shape of the initial ground surface.

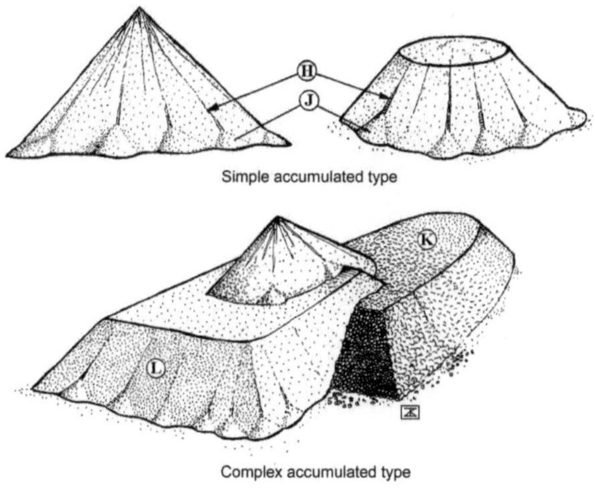

Fig. 9.4 Typical forms of quarry dumps: (**h**) gully, (**j**) debris cone, (**k**) plateau, (**l**) slope (Dávid and Karancsi 1999)

The most obvious *microforms* on dumps, formed by natural processes, are gullies cut into slopes. They are arranged radially on cone-shaped or truncated cone-shaped dumps. The dump material carried by rainwater settles in small alluvial cones at footslopes. Flat-topped plateaux may be dissected by headward eroding gullies.

The accumulated microforms of quarry floors, formed as a result of quarrying, are larger heaps and boulders dissecting the approximately flat ground surface.

9.2.4 Planation Activity

Quarrying does not only construct landforms but it can also result in *planation*. With the spreading of dump material over natural or artificial features (slopes, valleys, pits or depressions), they may be filled. Another possibility is the excavation of whole mountains during quarrying activities, resulting in huge landscape scars. Remarkable instances are found in Hungary (the Naszály at Vác, the Bélkő at Bélapátfalva (Plate 9.1), the Esztramos at Tornaszentandrás, the limestone quarries at the Szársomlyó Hills of Nagyharsány, the rhyolite quarry of the Kis Hill at Gyöngyössolymos, the laccolite of the Csódi Hill at Dunabogdány).

Plate 9.1 Excavation of the Bélkő near Bélapátfalva (source: http://www.pihenek.hu/krisztina_apartman/belapatfalvai_szallas/krisztina_apartman18.jpg)

9.3 Other Raw Materials for Construction

The utilisation of one of the most important building materials, i.e. the gravel, also goes back to a thousand years. Gravel was used for building and road constructions,

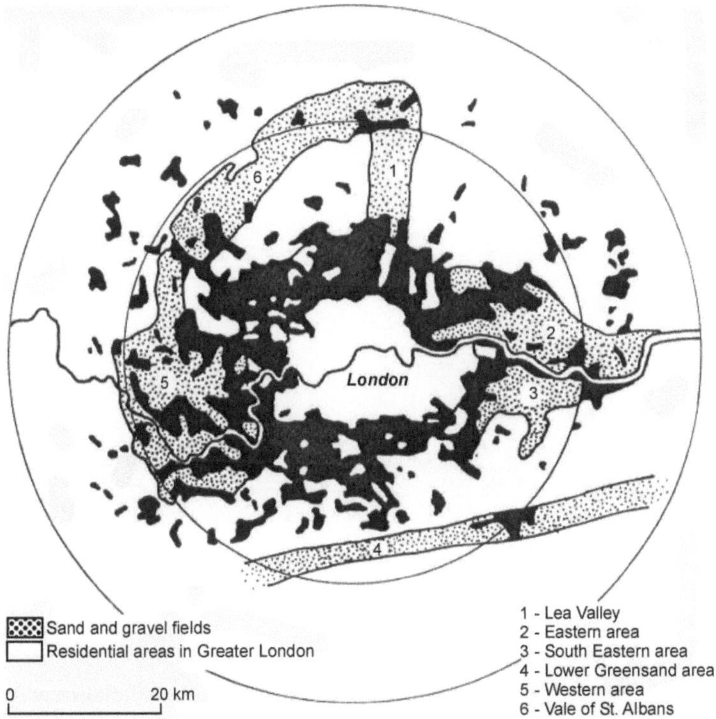

Fig. 9.5 Relevant sand and gravel excavation sites in Greater London (Nir 1983)

and accordingly, excavation sites were placed at the surroundings of settlements (Fig. 9.5).

Gravel extraction for construction purposes started in Hungary at the end of the 19th century, which was first restricted to the easily accessible gravel in the river beds exploitable by manpower. The first cement mill was established in the 1880s in the area of Lábatlan and Nyergesújfalu; since then the rate of gravel extraction used for constructions had been increasing. The material was obtained mainly by hand power; however, between the 1920s and 1940s, mechanic excavators were already applied. With the development of technical tools, underwater gravel exploitation also became possible. In the period between the two world wars, most of the gravel was extracted by dredging rivers; however, the extraction of other gravel deposits also started. This came to the fore when the use of concrete began to spread. At this time, extraction began in regions where gravel extraction is still typical today (the Alpokalja Region, watercourses of western Hungary, the Rábaköz and along the rivers Rába and Marcal, the Tapolcai Basin, the areas along the River Danube, e.g. Csepel, Délegyháza, the Mura and Dráva regions, the border section of the rivers Körös and Maros, forelands of the North Hungarian mountains, gravel sheets of the rivers Zagyva, Eger, Sajó and Hernád).

Among the surface landforms resulting from mining, the most common and typical ones are mine ponds accumulated from groundwater as well as waste dumps

Plate 9.2 Pond in gravel pit at Sajószöged (source: http://www.sajoszoged.hu/kepek.php?kep=Picture%20060.jpg)

piled up of the waste material (Plate 9.2). Their natural or artificial transformation begins almost immediately following their formation. Natural processes destroying both positive and negative landforms contribute to planation. The other factor is human action adjusting landforms to the purposes of human society. These two impacts can only be distinguished when a mining area is abandoned and natural processes take over its further shaping.

Landforms developed during mining activity are influenced by the same processes as natural landforms. On the sides of both negative and positive forms, various mass movements are present. Waste heaps of loose material and with steep sides, temporarily saturated, undercut bank sections or slopes may provide especially favour such processes. On steep slopes of loose materials (waste heaps, lake shores), rainwater runoff may induce erosion; on waste heaps radial networks of gullies can develop. By this, rainwater can wash (sometimes hazardous) substances from the shores or the heaps into the pond. On barren surfaces deflation can be also significant; however, vegetation on the slopes can decrease its impact.

On the shores of more extensive mine lakes, wave motion can play a relevant role. When steep failures occur, shore sections resulting from mining are abandoned; they are gradually transformed into slopes with stable angle, so-called self-adjusting slopes. In addition to this, wave motions also assort deposits from the lake bottom by depositing larger particles near the shore while carrying fine-grained substances to the central parts of the lake. In the transition of landforms resulting from mining, apart from natural processes, the operator or owner of the mine as well as residents of the surrounding area can also be instrumental. This impact can be diverse including unplanned spontaneous conversions and well-organised landscape architecture.

The aim of the latter is the reclamation of areas with limited utilisation possibilities due to mining activity, making them suitable for re-utilisation and fitting them into their environment. As during gravel excavation, a mine lake is evolved in most cases, by reclamation in a narrow sense, i.e. the total re-establishment of the original conditions cannot be the aim. In such cases, ways of alternative utilisation, for instance, for the purposes of tourism, are often envisaged.

Among building materials, the quarrying of sand, silica sand and clay must also be mentioned. As far as their landscape-forming impacts are concerned, they produce mostly negative landforms. At the outskirts of many settlements in the Great Hungarian Plain, former loam pits are still determinant elements of the landscape (with ponds, which unfortunately usually function as waste disposal sites).

9.4 Other Minerals

Among *other mineral stocks*, the most spectacular surface landforms are produced by the mining of various ores (in Hungary, bauxite, iron, uranium and manganese ores and sulphides), as well as the quarrying of perlite, bentonite and other compound minerals. Some scenic examples are going to be introduced below by the mining of rock salt, some ores and diamond. During their excavation, vast amounts of waste are generated compared to which the quantity of the target material of mining can be often negligible (Table 9.2).

In the case of rock salt, the focus should be on vast underground tunnels and halls (Marosújvár, Torda, Parajd, Aknaszlatina, Wieliczka), which later are used for the purposes of tourism and therapy (Plate 9.3).

The Bingham Canyon copper mine in Utah (USA) (Plate 9.4 – with an area of 5.4 km^2 and a depth of 695 m), the great vent of Kimberley, the silent witness of the diamond boom in South Africa (Plate 9.5) or the diamond mine near Mirny in Siberia (Plate 9.6 – with a diameter of 1.5 km and a depth of 540 m) are remarkable. One of the iron-ore pits of the Kursk Magnetic Anomaly is 7 km in length, 3 km in width and 90 m in depth (Mihajlovka). Special mining techniques

Table 9.2 Waste generation by the mining of the six raw materials most important in this respect (Nir 1983)

Waste generated by the quarrying of raw materials in the world (1977)	
Material	Material product gained as percentage of the material moved
Iron	50
Bauxite	55
Gold	0.9×10^{-3}
Copper	2
Coal	50
Brown coal	50

Plate 9.3 The great hall of the salt mine in Parajd (Transylvania) (source: http://s021.yatko.com/~kukullo/imagebank/csillagturak/sovidek/parajd2.jpg)

Plate 9.4 The Bingham Canyon copper mine in Utah, USA (source: http://static.panoramio.com/photos/original/5101732.jpg)

Plate 9.5 Vent of the great diamond mine of Kimberley in South Africa (source: http://www.gps.caltech.edu/news/features/southafrica/pics/d13bighole_mb.jpg)

Plate 9.6 Diamond mine near Mirny in Siberia (source: http://www.acrosstheplanet.net/wp-content/uploads/2008/05/mirny1.jpg)

may result in unique and, from a cultural historical point of view, valuable landforms (Recsk, Gyöngyösoroszi, the mining of veins and placers in Telkibánya (Hungary), or landforms left behind by gold-diggers in Ballarat, Australia, where a thematic park has also been constructed in the derelict mine).

9.5 Some Problems of Mine Opening and After-use

Mine openings and developments are often attended by vociferous opposition. Today's best example is given by the plans of gold mining near Verespatak (Rosia Montana) in Transylvania (Plate 9.7) where, according to estimates, to get gold for a golden ring, 20 tonnes of mining waste would have to be excavated. The mountains around Verespatak contain 300 tonnes of gold and 1600 tonnes of silver, making it one of the richest gold deposits of the world. To have it excavated, 225 millions of tonnes of rocks must be processed and 20 tonnes of dynamite would have to be exploded each day throughout 15 years.

Until recently, *abandoned quarries* both in Hungary and abroad have raised negative, unpleasant associations as 'scars in the landscape' (Plate 9.8). However, according to a new assessment, abandoned quarries and mines are regarded as 'environmental values', appreciated as possible sites for various uses (Bauer 1970; Dávid and Patrick 1999), i.e. exhibition sites or scenes for regional and tourism development projects (e.g. Fertőrákos, Ság Hill, the Kálvária Hill of Tata). The case studies below intend to provide a brief overview.

Plate 9.7 The mine of Verespatak (source: http://storage0.dms.mpinteractiv.ro/media/1/186/6450/2378245/1/rosia-montana-vt.jpg)

Plate 9.8 Scar of the Sás-tó andesite quarry near Gyöngyös made visible by a clear-cut in the Mátra Mountains, with the peak Kékes-tető (1014 m) in the background (Dávid 2004)

9.5.1 Case Studies

9.5.1.1 Rehabilitation of Abandoned Mining Grounds in Cornwall (England)

Cornwall, located at the south western part of England, is a region formed by settlements traditionally involved in the mining of noble ores. It was a centre of copper and tin mining from the 16th century until the end of the last century. Copper and tin exploration has also left significant environmental damage behind in the region. During the time of this project, more than ten mining grounds were reclaimed and a further five are at the stage of planning. At these grounds, 31 buildings were secured and further 20 are expected to be rescued. The historical scenery and unique atmosphere provided by old mine tunnels, dressing-rooms and engine-houses is intended to be preserved as well.

Along the track of the railway connecting the mines, tourist paths and nature trails were established. The path network will total 100 km and, in addition to providing ideal possibilities for tracking, cycling and horse-riding, it will also connect the settlements for the benefit of local residents. Wandering along these routes, several mining sites, machine-houses and other buildings converted into museums can be seen. This mining landscape represents such a special and unique value that an application was prepared (Fekete 2001) and in 2007 it was declared to be a World Cultural Heritage Site.

9.5.1.2 Bluewater Shopping Centre

In recent years, many precedents, mainly from Great Britain show that commercial centres (hyper- and supermarkets) are constructed in old quarries outside cities and next to them facilities for entertainment (parks, multiplex cinemas, gaming-rooms,

Plate 9.9 The Bluewater shopping centre near London with the wall of the blue circle chalk quarry in the background (source: http://upload.wikimedia.org/wikipedia/commons/7/7e/Bluewater_Shopping_Centre,_Kent,_England_Crop_-_April_2009.jpg)

concert halls, discos, galleries, art centres, etc.) are also developed (Bennett and Doyle 1999). The most outstanding example for this is the Blue Water Shopping Centre located in Dartford at Junction No. 2 of the London Ring Road M25, marketed as the largest entertainment centre of this kind in Europe. This investment, compelling both in its outside and inside appearance, has been built between 1995 and 1999, in the area of the abandoned Blue Circle Chalk Quarry (Plate 9.9).

9.5.1.3 Tokaj–Patkó Quarry

The former quarry hosted a large-scale cultural event on 30th June 2002, functioning as a 'festival cauldron'. The event took place on the occasion that the Tokaj-Hegyalja Region was awarded the UNESCO World Heritage status in the category

Plate 9.10 Concert and festival in the Patkó Quarry at Tokaj (30th June 2002) (source: http://www.tokaj.hu/galeria_eletkepek)

Plate 9.11 Sculpture park in the Hársas Hill of Nagyharsány (source: http://upload.wikimedia.org/wikipedia/commons/a/a8/Nagyharsanyi_szoborpark.jpg)

of cultural landscapes. Cultural programs were organised in the quarry to celebrate it (Plate 9.10). Since then it has been regularly used as a site for similar events.

9.5.1.4 Sculpture Park in the Hársas Hill of Nagyharsány

This quarry is a good example of after-use for the purposes of fine-arts. The excavation of limestone for cement industry had already begun at the Eastern side of the hill in the early 20th century (Lóczy et al. 2007); a sculpture park can be found in the former quarry floor (Plate 9.11).

References

Bauer AM (1970) A Guide to Site Development Rehabilitation of Pits and Quarries. IMR 33. Ontario Department of Mines, Toronto, Canada

Bennett MR, Doyle P (1999) Environmental Geology. John Wiley and Sons, Chichester

Dávid L (2000) A kőbányászat, mint felszínalakító tevékenység tájvédelmi, tájrendezési és területfejlesztési vonatkozásai Mátra-hegységi példák alapján (Landscape Protection, Landscape Management and Regional Development Aspects of Quarrying as a Surface-Forming Activity Based on Examples from the Mátra Mountains). PhD Thesis. Debreceni Egyetem, Debrecen

Dávid L, Baros Z (2006) Impacts of Tourism and Sport Activities on the Surface of the Earth – An Anthropogenic Geomorphological Approach. e-Terra, GEOTIC – Gesciences On-Line Journal, Portuguese Geological Society Group, Portugal,(http://e-terra.geopor.pt), 4(1)

Dávid L, Karancsi Z (1999) Analysis of Anthropogenic Effects of Quarries in a Hungarian Basalt Volcanic Area. 2nd International Conference of PhD Students, University of Miskolc, Miskolc

Dávid L, Patrick C (1998) Quarrying as an anthropogenic geomorphological activity. In: Szabó J, Wach J (Eds.), Anthropogenic Aspects of Geographical Environment Transformation. University of Silesia, Faculty of Earth Sciences,Sosnowiec, Kossuth Lajos University, Department of Physical Geography, Debrecen, Debrecen-Sosnowiec

Dávid L, Patrick C (1999) Problems of Quarry Restoration: General Considerations and Some Examples from North-West England and Hungary. In: Modern Nature Use and Anthropogenic Processes. University of Silesia, Faculty of Earth Sciences, Sosnowiec, Academy of Sciences, Irkutsk

Erdősi F (1966) A bányászat felszínformáló jelentősége (The relevance of mining in surface evolution). Földr Közl 90 (4): 324–343

Erdősi F (1969) Az antropogén geomorfológia mint új földrajzi tudományág (Anthropogenic geomorphology as a new discipline of geography). Földr Közl 93 (1): 11–26

Erdősi F (1987) A társadalom hatása a felszínre, a vizekre és az éghajlatra a Mecsek tágabb környezetében (Impact of Society on the Surface, Water and Climate in the Broader Environs of Mecsek Mountains). Akadémiai Kiadó, Budapest

Fekete É (2001) Együtt – De hogyan?," Innovációk a kistérségi fejlesztésekben ('Together – But How?' Innovations in the development of Micro-Regions). MTA RKK, Miskolc-Pécs

Karancsi Z (2000) A kőbányászat során kialakult felszínformák tipizálása (Classification of Landforms Resultant from Quarrying). In: A táj és az ember – geográfus szemmel (Geográfus doktoranduszok IV. országos konferenciája, 22nd–23rd October 1999). Szeged. http://phd.ini.hu

Lóczy D, Gyenizse P, Ronczyk L, Dezső J (2007) Damage to the landscape caused by cement industry in Southern Transdanubia, Hungary. Geogr Fis Dinam Quat 30 (2): 209–213

Nir D (1983) Man, A Geomorphological Agent. Keter Publishing House, Jerusalem, Boston, London

Ozorai Gy (1955) A kőbányászat kézikönyve I-II. (A Handbook of Quarrying Vol. 1–2). Műszaki Könyvkiadó, Budapest

Simán K (1995) Catalogue of flint mines: Hungary. Archaeol Polona 33: 370–382

Website Sources

http://www.pihenek.hu/krisztina_apartman/belapatfalvai_szallas/krisztina_apartman18.jpg
http://www.sajoszoged.hu/kepek.php?kep=Picture%20060.jpg
http://www.s021.yatko.com/~kukullo/imagebank/csillagturak/sovidek/parajd2.jpg
http://static.panoramio.com/photos/original/5101732.jpg
http://www.gps.caltech.edu/news/features/southafrica/pics/d13bighole_mb.jpg
http://www.acrosstheplanet.net/wp-content/uploads/2008/05/mirny1.jpg
http://storage0.dms.mpinteractiv.ro/media/1/186/6450/2378245/1/rosia-montana-vt.jpg
http://www.haldesign.hu/ghaleria/images/2007/200.jpg
http://upload.wikimedia.org/wikipedia/commons/7/7e/Bluewater_Shopping_Centre,_Kent,_England_Crop_-_April_2009.jpg
http://www.tokaj.hu/galeria_eletkepek
http://upload.wikimedia.org/wikipedia/commons/a/a8/Nagyharsanyi_szoborpark.jpg

Chapter 10
Mining: Extraction of Fossil Fuels

László Sütő

Abstract The establishment of opencast mines is a kind of direct excavational process. From a geomorphological aspect, after the removal of the overburden and during working the seams, erosion starts immediately, but is kept under control to secure exploitation during the time of extraction. In this case, the stabilisation of the side-walls and the drainage of spring water are problems to be solved. Vertical dislocations, called subsidence, are semi-anthropogenic processes. Major subsidence landforms develop after the termination of uniform mechanised extraction, when large subsurface hollows of several metres' height collapse. Large-scale surface alterations are observed between a few days and a few years following mine closure. Water saturation in the fractured zones of loose rock brings about specific morphological changes, even rises in surface elevation. As a by-product of coal mining, inert overlying sediments and the intermediate waste rock between coal layers are piled in the form of spoil heaps. These hills with a height of more than 100 m are hallmarks even after landscaping. The extraction of mineral oil and natural gas also result in modification of topography in the region affected.

Keywords Opencast coal mining · Deep mining · Subsidence · Waste tips · Fluid fuels

10.1 Introduction

Fossil raw materials and fuels are essential to the sustenance of humans since the Palaeolithic times. Their discovery is considered to be a driving force of social development. At the beginning, no attention was paid to the resulting scars in the

L. Sütő (✉)
Department of Tourism and Geography, College of Nyíregyháza, 4401 Nyíregyháza, Hungary
e-mail: sutolaci@nyf.hu

J. Szabó et al. (eds.), *Anthropogenic Geomorphology*,
DOI 10.1007/978-90-481-3058-0_10, © Springer Science+Business Media B.V. 2010

landscape as their extent seemed to be minimal and after being abandoned, they soon adjusted to the landscape through natural geomorphic processes.

The mining of fuels started by the technical revolution, however, altered the landscape itself, and by today, the reworking of surface materials by humans – as discussed in former chapters – in many fields exceeded the impact of natural processes (Sherlock 1922; Nir 1981; Erdősi 1987; Pécsi 1991; Szabó 1993). According to Sherlock (1922), the amount of material moved by mining in Great Britain amounts to 88% of the total material reworked. Between 1900 and 1975, earth removal related to underground mining alone is estimated to be 110 km^3 (Louis and Fischer 1979). Thus, in sense of intensity and spatial extension, mining is regarded to be one of the most effective processes.

The environments where fossil fuels form and accumulate are diverse; consequently, so are their properties. The duration of the anthropogenic processes active during their extraction as well as the modified landforms are also variable. *Montanogenic landforms* related to the mining of fossil fuels including excavation, levelling and accumulation activities are, in all cases, the *results of direct secondary processes*, as they are not intended goals of mining but only its by-products. Aggrading concave landforms, degrading convex landforms and neutral flat terrains in quasi-equilibrium are formed. On these newly developed surfaces, natural *denudation processes* under the given climatic conditions will produce *secondary* (*semi-anthropogenic* or *natural-anthropogenic*) *processes landforms* (Erdősi 1987; Szabó 1993).

10.2 Surface Alterations Caused by Opencast Coal Mining

The establishment of opencast mines is a kind of direct excavational process. For the excavation of the brown coal sequence of 70 m thickness near Cologne, over an area of 80 km^2, a sediment cover locally of 450 m thickness has to be removed (Szabó 1993). If a truck were used in an opencast mine of this size to transport coal (as it is done in the case of more valuable raw materials), it would not even reach the mine floor within 1 hour. *Indirect ground subsidence* resultant from the lowering of water table in opencast mines may affect an area much larger than that being mined.

From a geomorphological aspect, after the removal of the overburden and during working the seams, erosion starts immediately. It is kept under control, however, to secure exploitation during the time of extraction. In this case, the stabilisation of the side-walls and the drainage of spring water represent problems to be solved.

After the opening of the opencast mine at Ecséd in 1957 and at Visonta in 1969, water extraction induced severe ground subsidence with serious damage to buildings in the nearby settlements (Dávid 1999). In the lignite mining area at the foothills of the Mátra Mountains, deep mines were operated until the 1960s, but major surface movements undoubtedly resulted from water extraction in the environs of opencast pits (Fig. 10.1).

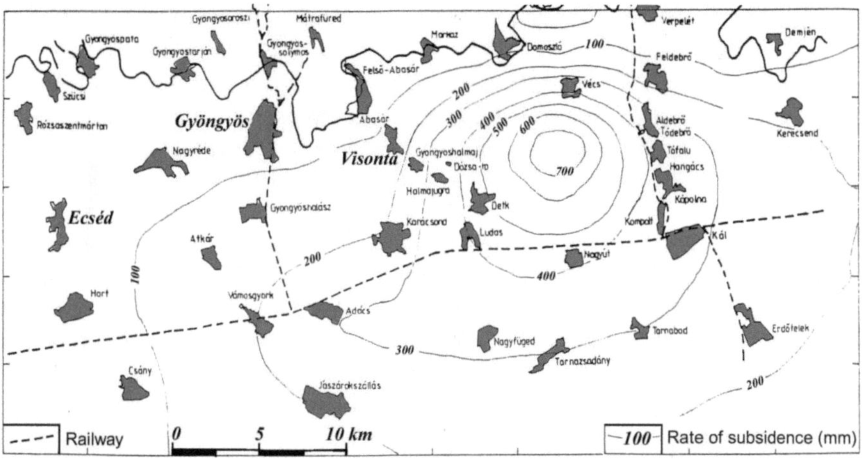

Fig. 10.1 The rate of ground subsidence due to lignite mining at Ecséd and Visonta (after Jambrik 1993)

In the mining areas, still continuously monitored, a concentric depression has formed with a maximum depth of 70 cm in 2000. Investigations indicate that, in addition to a water table depression of 35–70 m, the inflexion point of the subsidence appeared 5–7 years after the beginning of movements, and continues with gradually decreasing intensity when the pit is backfilled. The close connection between water table lowering and surface movements is well indicated by the fact that backfilling only moderately raised the surface when opencast mining was recently extended to the area around the villages of Visonta and Abasár, and at the nearby village of Halmajugra, further subsidence occurred (Jambrik 1993).

When opencast pits are abandoned, *mass movements* and *gullying* occur on steep slopes previously maintained. Underground water rising to the initial water level can often fill up the artificially created basin as a pond. Its size is limited if during subsequent landscaping waste material, with the exception of the final void, is backfilled into the pit. However, a pond or a cluster of ponds functioning as *local base level* can form. It may be favourable in arid regions but may have uncertain impacts over the long term (Kerényi 1995).

When environmental aspects are taken into account, the site of the temporary waste tip is designated prior to mining, and the biologically valuable humic topsoil is stored in piles. The inert material, according to the principle of continuous recultivation, can be backfilled into the excavated open pit and topsoiled at the completion of the mining activity (Kerényi 1995). The backfilled terrain will be somewhat lower compared to the initial surface and the unconsolidated waste may gradually compact to a significant degree. In the meantime, the extension of the mining landscape with backfilling, surface evolution processes become localised and controlled by landscaping. Where deep and opencast mining are practised next to each other, the open pit can be filled up by waste material from the former, moderating environmental pressure. Major pits are capable of accommodating several tens of millions of cubic metres of waste material.

Ponds in opencast mines begin to fill up as soon as they are created. In the densely populated European mining regions, the environs of ponds are often transformed by human action. By spreading the collected waste and levelling the surface, the lake basin is stabilised, resorts are developed (Gefferth 1986). Contrasting developments between natural and artificially landscaped terrains can be studied in the Borsod Coal-Mining Area, between the opencast mines at Vadna (at the confluence of Sajó and Bán streams) and at Herbolya (south of the town of Kazincbarcika). The pond of 19 ha area in the former pit is a resort area. Sediments could be washed in into the entirely closed basin primarily from the spread waste material, but it has been prevented by the stepped piling of the excavated waste and the stabilisation of slopes. Consequently, only minor amounts of material are transported by the winds blowing in the Sajó Valley into the pond. Steep side-walls of mining origin were flattened to preclude major mass movements. Therefore, upfilling is only moderate until the new recreation function is maintained. On the contrary, the opencast mine at Herbolya, with a smaller area of 2.5 ha, follows a natural evolution and, therefore, slow mass redeposition from the nearby waste tips leads to upfilling. Its rate is indicated by the increasingly spreading shore vegetation.

10.3 Geomorphic Processes Related to Deep Mining of Coal

Deep mining of coal is characterised by *three types of primary landscape alterations*. Direct secondary landforms include *levelled terrains*, *mine workings* as excavation landforms and *waste tips* as accumulation landforms – all with various further evolutions. Mining-related landscape evolution is marked by road-cuts, supporting walls and embankments often cutting through beds with varying stability and water storage capacity. Supporting walls hindering material transport become cracked under stress. Erosion rills are often formed next to concrete castings, road-cuts can collapse, embankments can be underwashed and roads covered with debris. After the abandonment of galleries, subsidence develops and waste tips are affected by denudation.

10.3.1 Landforms of Mine Workings and Subsidence Areas

At the beginning of working, *headings* are deepened to gain access to coal deposits. Depending on the means of exploitation, workings to the coal deposits spring from the main headings, during the extraction of which, *underground passage networks* of various sizes are developed in accordance with local conditions.

In the 'golden age' of coal mining (in Hungary from the mid-18th to the mid-20th centuries) manual room-and-pillar working with wooden supports was applied (Benke and Reményi 1996). Mines were mostly simply abandoned; stowing was rarely employed (e.g. at Dorog or at Pereces near Miskolc). From the beginning, special attention was paid to more severe secondary surface movements to avoid disaster, damage prevention, however, was only practised after alterations. With developing technology, in most European mines, longwall workings with self-advancing support began to be used after World War II to obtain more coal. This technology, however, led to the formation of more extensive interconnected passage networks.

Underground hollows can be traced on maps of mines. As an example, the extension of workings, the location of transport and air shafts as well as headings in a mining area (around Lyukóbánya and Pereces shown in Fig. 10.2).

In the deep brown coal mining districts operated between 1830 and 2004 deposits were worked by both traditional and modern mining technologies. The total extension of undermined sections was approximately 21 million m^2 (Benke and Reményi 1996). Having this evenly divided over the two catchment areas, at an average depth of 100 m, it would equal a hollow with an average height of 3 m. But more than half of the region is actually undermined, and the area of mining impact is more than tenfold larger.

Approximately 50 km of *mine workings* were established here; this equals the total length of underground lines in Budapest. No nationwide data is available on the collapse of steel supports used from the late 1950s, as they theoretically resist

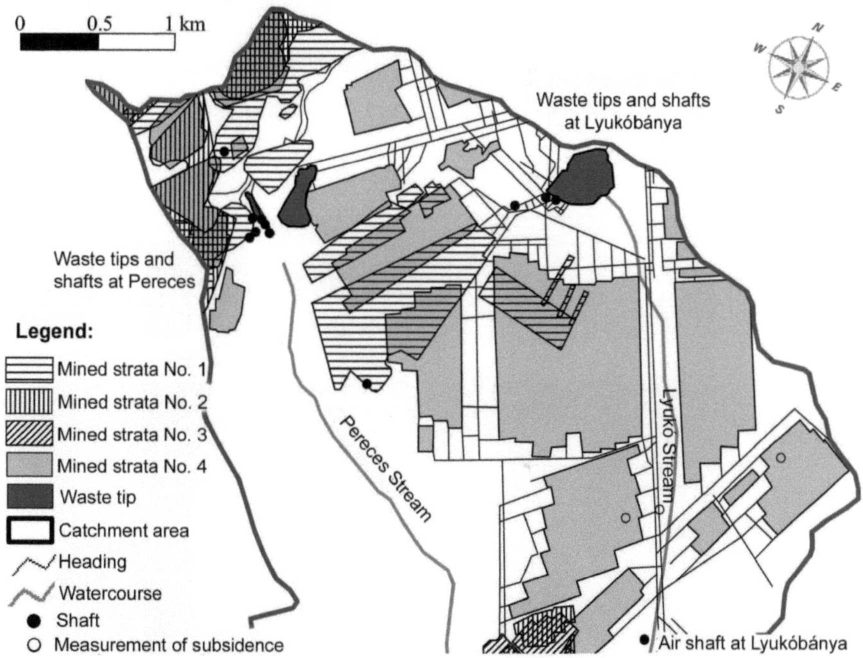

Fig. 10.2 A section of the simplified, digitised map of mines in the Lyukóbánya and Pereces region (Sütő et al. 2004)

the pressure of the overlying rock mass for hundreds of years. The deformations of security rings in abandoned mine workings, however, indicate movements (Plate 10.1). We can conclude that a secondary maximum occurs within one or two centuries caused by the overageing of mine working supports and subsidence models are unable to take it into account. Their manifestation on the surface, however, is associated with an excavated surface of critical size (Hoványi and Kolozsvári 1989). To avoid this, a concrete frame is applied in shafts (Plate 10.2). In hilly terrains

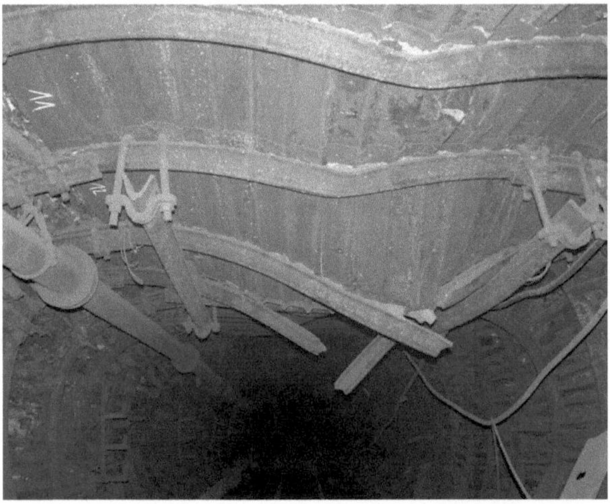

Plate 10.1 Deformation of a TH4 steel ring used for mine working support after three decades (Sütő 2003)

Plate 10.2 A shaft section of the main pit in Lyukóbánya secured by a concrete frame and steel rings (Sütő 2003)

dissected by erosion valleys with brown coal deposits, however, when incision reaches a critical depth, stepped failure faces are formed.

Vertical dislocations, called *subsidence*, are semi-anthropogenic processes (Ádám and Pécsi 1985; Fodor and Kleb 1986). The form and scale of surface movements triggered by the collapse of hollows depend on the duration of excavation, on the spatial extension and thickness of the excavated strata, their depth under ground, structural and hydro-geologic conditions, the rock mechanics and other lithological features of the overlying strata and on the mining techniques applied (excavation method, support, abandonment) (Martos 1956, 1958; Erdősi 1987). The last mentioned two groups of factors will then determine how much time is required for subsidence to occur on the surface (Hoványi and Kolozsvári 1989). Various empirical formulas were developed to estimate the *extension and rate* of subsidence (Martos 1955, 1956; Nir 1981; Hoványi and Kolozsvári 1989; Somosvári 1989, 1990).

Major subsidence landforms develop after the termination of uniform mechanised extraction, when large subsurface hollows of several metres' height collapse. Large-scale surface alterations are observed between a few days and a few years following the end of mining (Erdősi 1987; Ládai 2002). The period of time when the rate of monthly subsidence is between 30 and 50 mm is called the active phase. In a mine of the Mecsek Mountains following the extraction of a deposit at 300 m depth, major surface movements at a rate of 3 m per day, took place in about 3–4 months (Erdősi 1987). In the decaying phase, debris compaction is predominant and, consequently, the intensity and the rate of subsidence will be lower (Ládai 2002). Movements are regarded inactive if displacements less than 30 mm in a half-year period are observed (Hoványi and Kolozsvári 1989). However, subsidence eventually ceases much later. In the Northern mine of Nagyegyháza, 99.9% of the entire subsidence took place within 8 years after extraction. For the Mány mine, 99.9% of the entire subsidence is estimated to happen within 12 years after mine closure (Ládai 2002).

Subsidence profiles mostly resemble a Gaussian distribution curve adjusted to various environmental conditions (Fig. 10.3). In the case of horizontal beds, maximum depth will appear above the centre of the excavated area (Martos 1955). Above subsidence, two zones can be distinguished: a zone of compression with mass movements and a zone of expansion with fractures and graben-like subsidence (Erdősi 1987; Hoványi and Kolozsvári 1989). Horizontal displacements are typically directed inward the concave landform. According to the measurements by Martos (1956), they already occur in the inflexion point of the profile curve; in the coal mines of the Transdanubian Mountains they reached 65 cm with a maximum daily rate of 2 cm a month after extraction.

The horizontal extension of subsidence is calculated from the angle of draw, i.e. the angle to the horizontal of a straight line drawn between the margin of the mine hollow and a stable external point (Martos 1955). The angle of draw is an acute angle pointing outward from the site, therefore the surface projection of the mine working is always larger than the area directly undermined (Martos 1955; Erdősi 1987; Hoványi and Kolozsvári 1989) (Figure 10.3). The area affected by subsidence is calculated from the formula

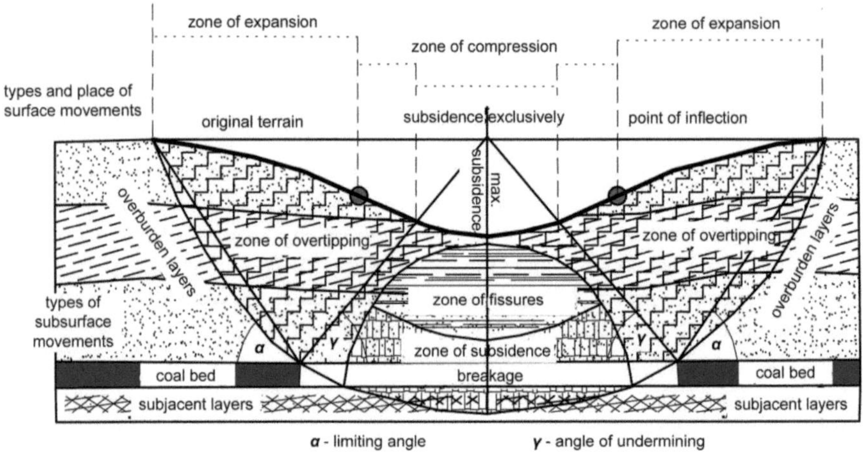

Fig. 10.3 Sketch of subsurface and surface subsidence elements (modified after Erdősi 1987; Hoványi and Kolozsvári 1989)

$$r = H \times \cot\beta \tag{10.1}$$

(where H: thickness of heading, β: the angle of draw, being 46° for the Lyukó deposit No. 4, r: impact radius)

The result of the calculation for Lyukóbánya is 260 m. This is the distance from the surface projection of the working at which surface movements are observed – although with attenuation.

The surface extension of subsidence is influenced by several factors. If two deposits of similar environmental conditions are worked, ground subsidence of a greater extent will occur for the deeper-lying deposit; however, in this case, the depression produced will be shallow and bowl-shaped, whereas for shallower deposits, deeper bowl-shaped crown holes with shorter radius are expected (Hoványi and Kolozsvári 1989).

The physical features of the embedding *waste rock* also induce various impacts. On the unconsolidated country rocks of young coal deposits, by steady wave-like dispersion, broader surface depressions form and spread radially like waves often within weeks. In the case of compact waste rocks, movements take place with a higher angle of draw. Thus, minor depressions evolving slowly and in stages are expected. In the meantime, in the 50–100 m zone of intense tectonics, the rate of surface displacements is increased and their duration is shortened because of rock mechanics properties (Hoványi and Kolozsvári 1989). On slopes, asymmetric depressions form with their maximum depth along the axis of the mining area, where the deposits lie in the highest position (Martos 1956). When steeper slope is associated with unconsolidated footwall, displacement extends over the latter and surface movements intensify.

When several *superimposed seams* are mined, the impact of displacement is higher on the already loosened terrain. Angle of draw may drop by 5–10° resulting in significantly wider surface impact radius (Hoványi and Kolozsvári 1989). Based on the number of superimposed mined seams, various *morphological districts* can be identified. In the northern part of the mining area at Pereces, three superimposed seams were mined (Figure 10.2) and failure faces, gullies and landslides resulted. In the northern part of Lyukóbánya, two seams were mined, but at more than 200 m depth on average and, consequently, no remarkable topographical alteration was generated. But, as a result of the less favourable geological settings, among the surface forming processes, slides predominate and mining could contribute to their re-activation. Surface movements in the southern part present an environmental hazard of high vulnerability. Although the surface distance of the worked seams is usually only 50–100 m, they are located next to Miskolc, a city with a densely built-up area. Locally, waste tips also represent a surplus load.

As a result of subsidence, depressions, ruptures, cracks and deformations developed on the surface fall into various size scales. Classified by Erdősi (1987), they are arranged in order of magnitude: subsidence trough > depression field > slumping > dell (dry valley).

A *subsidence trough* is composed of depressions in mining areas, which are gradually merging over a period of more than a decade. In the basically shallow depression field with an area of several km^2, depressions of various sizes on the diverse micro-landforms (fault scarps, minor dells, etc.) and, above the pillars, zones hardly affected by movements also occur. Based on findings in Hungary, the deepest subsidence troughs are deeper than 20 m with an affected area of more than 10 km^2. Compared to the area undermined, the size of the resultant depression can be tenfold.

In the Mecsek Mountains the subsidence trough at Pécs-Somogy is the largest mesoform (area: 13.5 km^2, maximum depth 27 m), covering sevenfold of the undermined area. The formation of this trough resulted in a topographic inversion; its deeper sections may have been formed at former heights. Among the fault scarps dissecting the surface, a remarkable dislocation developed at the mined surface projection of the István shaft at Pécs, 370 m in length and with a throw of 1.8 m. This old scarp can be distinguished from natural failure faces or abandoned vineyard terraces (Erdősi 1987).

A *depression field* is a larger uninterrupted feature formed within a trough, whose microfeatures are adjusted to tensile and compressional forces. An example is found in the area of the István Shaft of Pécs with a rectangular groundplan and an area of 106 m^2, a depth of 6.8 m and a gradient of 32–45°. Where tensile force comes to its maximum, in the case of more solid rocks, regular faults are also formed, as on Juhász Hill at Nagymányok in the Mecsek Mountains where swallow hole-like features occurred (Erdősi 1987.)

Collapses are steep-sided, smaller pits with a diameter of 10–20 m. They are usually found at a few tens of metresdistance from shafts of near-surface seams. Such landforms with an average depth and diameter of 1–5 m studied in the Királd mining area of the Borsod Coal-Mining Region indicate a regular distribution along a line, exactly coinciding with the surface projection of excavations.

A *crown hole* is a symmetric collapse depression resultant from the mining of a single seam (Martos 1955). Crown holes are classified according to vertical motions, overburden thickness and excavation height (Figure 10.3). Right above the seams, in the collapse zone, overlying strata are totally cracked and loosened. The zone of cracked rock, although heavily jointed, still retains some coherence and withstands loading. Finally, the shelter zone follows with no jointing or loosening (Hoványi and Kolozsvári 1989). It is extremely important to delimit the collapse zone properly when below the mined seam, further excavation starts as the jointed and loosened overlying strata have a reduced load capacity.

When these zones become exposed on the surface *collapse*, a *fractured* and finally *displacement crown holes* form, based on the formula

$$H = m/k - 1 \tag{10.2}$$

(where H: overburden thickness; m: excavation height (with a maximum value of 4–5 m in the area of the Borsod Coal-Mining Area); k: a coefficient of loosening, its average value for Hungary is 1.2 for coal seams with a thickness of 3 m) (Hoványi and Kolozsvári 1989).

Crown holes form where the thickness of the overlying strata does not exceed 20–25 m. In the Borsod Coal-Mining Area, the mining of most coal seams began along the main tributary valleys to the Sajó River; thus, collapse crown holes developed here (Bertalanffy et al. 1986). The belt most impacted by mining coincides with the more unstable surfaces and slopes affected by slides. The formation of fractured crown holes, according to the formula (Hoványi and Kolozsvári 1989)

$$H < 40 - 50\,\mathrm{m} \tag{10.3}$$

is presumed in overlying deposits with a thickness less than 200–250 m. Smaller displacements also occur where overlying strata are thicker and crown holes are also found outside the depressions causing severe damage.

Water saturation in the fractured zones of loose rock brings about specific morphological changes. If accumulation increases pore pressure, vertical dilatation and occasionally *ground elevation* ensue. Based on investigations in the mining areas around the Mecsek Mountains, above multiple undermined areas where ground subsidence reached 20 m, the loosened zone could be 200–300 m thick, the depth of mining 400–600 m, whereas the surface was raised by more than 10 cm (Somosvári 2002).

With slope conditions altered by subsidence and surface levelling, the rate and intensity of *erosion-derasion processes* also change. Relative relief increased around the depression induces a more intensive regression of gullies. At Szászvár in the East Mecsek Mountains, gullies in the mining area incised 1–3 m in 15 years (Erdősi 1987).

Based on Erdősi (1987) and our own observations, depressions become rounded by erosion. Their slope drops to 10–15° and *accumulation* gradually reduces their depth to 1–5 m. Rock properties are also influential at this phase, as on loose sediments the processes tend to be more rapid. Modified slopes bring about a new base level in the sedimentary basin. This also receives sediments from the undisturbed terrain in the external impact area of the depression. Permanent watercourses accumulate in an uninterrupted manner; otherwise the accumulation by intermittent ravines and wind is predominant.

A good example of the total alteration of the topography and hydrology is mining around Tatabánya. By the 1980s, due to the collapsing method applied between the waterlogged depressions developed around Bánhida Lake, the watercourses formerly crossing the area intersected; a drainless area of continuous accumulation formed in the Által Stream floodplain. The largest depression in the mining area was occupied by a pond of more than 10,000 m^2 area and the flora of the region has also radically changed (Kleb 1990).

As a result of exogenic forces, the depression of István shaft in Pécs has expanded 50% in 6 years to 148 m^2; acquired a flattened, rounded, oval shape and has been accumulated to 2.6 m. A 2.3 m deep and 6 m wide hollow at a nearby system of gullies was filled up, in 8 years, by a sediment layer 80 cm in thickness while slope inclination reduced to its half (Erdősi 1987).

Similar processes took place in the Upper Silesian and Rybnik mining districts of the Katowice Voivodeship (Province), where the areas affected by ground subsidence due to mining, with depth exceeding 10–35 m, are estimated at 22% of the total area. Changes in relative relief the gradient of the Klodnyica River in Upper Silesia and created waterfalls along scarps in the ponded depressions. The waste material accumulated in artificial 'levèes' running along the valley and mass movements re-activated (Wach and Szczypek 1996).

10.3.2 Surface Evolution of Waste Tips

As a by-product of coal mining, inert overlying sediments and the intermediate waste rock between coal layers are piled in the form of *waste tips*. These hills of more than 100 m height are hallmarks even after landscaping. Therefore, they are regarded as decisive *accumulation landscape features* of mining origin. In Hungary, until the decline of coal mining, 248 waste tips of coal were accumulated, containing a total amount of 200 million m^3 material (Table 10.1).

The primary shape of waste tips depends on the rock mechanical properties and the amount of waste rock, on the topography of the disposal site and on the technology applied (Erdősi 1987; Szabó 1993; Sütő 2000, 2007), whereas their size is related to technological progress of mining. Manual room-and-pillar mining followed the bedding of coal seams almost precisely, and due to the use of human (or animal) power, reduced waste to the least possible amount. With the application of mechanised longwall mining and conveyance belt, mine working time shortened,

Table 10.1 The amount of waste material and the number of waste tips of coal mines in Hungary (Egerer and Namesánszky 1990)

Waste tips/County	Number	Size (million m^3)
1. Heves	13	59.1
2. Baranya	38	55.9
3. Komárom-Esztergom	26	26.1
4. Nógrád	76	23.4
5. Borsod-Abaúj-Zemplén	40	19.7

but a larger amount of waste, also containing more coal and organic matter, was excavated. Therefore, disposal sites were sought where land of smaller area and of economically less value had to be taken out of cultivation. Consequently, the waste material had to be deposited near pits. This way, however, unstable surfaces are affected by both ground subsidence over hollows and excess loading by waste tips. Therefore, the material of the initial terrain squeezed out from below the waste tip itself will suffer mass movements.

The *volume of waste tips* alone can be several million m^3. The total volume of the stepped double waste tip at the Béke and István shafts around Pécs is 11.2 million m^3, the maximum height is 67 m and occupies a triangle-shaped area 750 m long and 550 m wide (Erdősi 1987). To one of the largest waste tips of the Borsod Coal-Mining Area (in the Ádám Valley near Kazincbarcika), material was transported from several mines and coal separators between 1923 and 1988 (Bertalanffy et al. 1986). Now it is approximately 800 m long, 400 m wide, more than 100 m high and contains 1.2 million m^3 of material. Certain waste tips in the Upper Silesia industrial region of Poland accommodate sports fields of a housing estate or recreation park or even a railway track of over 1 km in length.

From the point of view of disposal, the initial relief can be negative, positive or neutral. Combining them with various disposal techniques, the following *types of waste tips* are distinguished (Fig. 10.4).

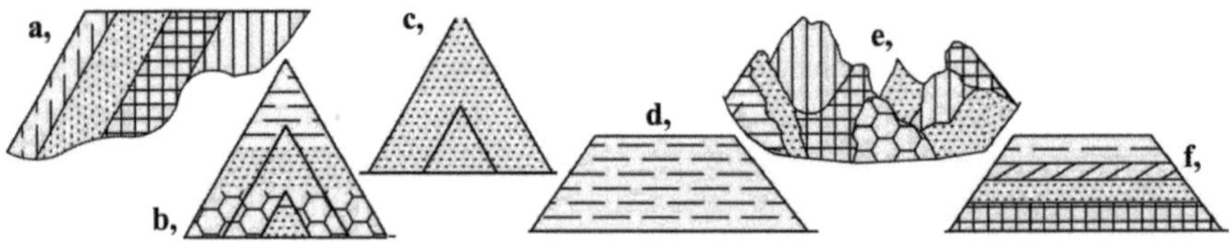

Fig. 10.4 Landforms and structure of waste tips based on the site of disposal and way of accumulation *(modified after Egerer and Namesánszky 1990)* (**a**) Slope-margin tip deposited on valley side with homogeneous grain-size distribution and inclined beds; (**b**) Doneck-type tip with layers of inhomogeneous grain-size distribution; (**c**) Doneck-type tip with homogeneous grain-size distribution; (**d**) Flat disposal on valley floor with homogeneous grain-size distribution; (**e**) Unlayered tip deposited at valley head and valley sides with inhomogeneous grain-size distribution; (**f**) Levelled flat disposal on valley floor with quasi-homogeneous grain-size distribution

During hydraulic conveyance, sludge-like material is placed on the tip. *Homogeneous* grain size is defined as required by the technical parameters of the conveyance system. A particular sub-type is represented by *sludge reservoirs* where the waste material or sludge in its plastic state mixed with fluid fills in some kind of ground depression. The resultant surfaces can rise more than 10 m above the initial surface, while the slow compaction of waste causes subsidence. A former lake basin is filled up by sludge on the margin of the Bánhida Lake at Tatabánya (Kleb 1990) or the settling tank of the Berente Thermal Power Plant at Kazincbarcika, located on the floodplain of the Sajó River. When the waste material comes from several mines or, for any other reason, the quality of the transported material varies, *layered tips* are accumulated. They are usually of trapezoid or of conical (Doneck type) shape and stable with well-designed slopes. The typical steep-sided waste tips remain features alien to the landscape even after reclamation.

The sorting of the piled material depends on disposal technology. At waste tips piled from coal tubs, gravitational sorting takes place, i.e. the coarser roll down to the foot of the slope. If transported by trucks or artificially levelled, waste is only partially sorted. The inhomogeneous grain-size distribution of tips of dipping layers or of those piled on flat terrains, makes them conical. Of waste piled juxtaposed from tubs along a cableway, wedges, ridges spotted by small peaks are formed.

When waste tips are established in the extension of valley sides, an excess mass is piled on positive landforms. In this case, a type called marginal tip by Erdősi (1987) is formed. It is an asymmetric landform with steep slopes only towards the valley, in accordance to the relief of the disposal surface – adjusted to the initial roughness of the surface, joined at the natural slope by a narrow summit. More often on inhomogeneous, non-stratified heaps, rarely on layered heaps with semi-homogeneous material, variation in rock mechanics can induce sliding and rockfalls until a new equilibrium is reached. Disposal features are less striking if a valley (valley head, side or floor) is filled up by inert material but it is most relevant for surface evolution. On the emerging new relief, the former equilibrium is upset (drainage is modified by valley impoundment), and geomorphic evolution takes a new trend. As the material of the natural surface and the mechanical properties of the waste rocks are different, the waste tip soaked by surface runoff can become unstable. Along the edge of the natural and yet unconsolidated waste material, a slip plane may form. In extreme cases *valleys* are entirely *filled up* by waste material, and this 'new hillside' is adjusted to the natural slope and the anthropogenic landform is only indicated by the flat top.

Natural geomorphic evolution processes on waste tips are more rapid than geological erosion in the area. Therefore, they can be regarded as short-term monitoring models (Homoki et al. 2000). Geomorphic evolution is closely correlated with the material of waste tips, weather changes, slope gradient and vegetation cover. Similarly to natural evolution, such processes tend to be more intense on steeper tips compared to initially flat-top landforms. Consequently, waste tips follow a similar path of evolution, i.e. flatten out and become more rounded.

On waste tips, from excess decomposition heat of bacterio-pyrite, *spontaneous combustion* and permanent *burning* of the partially siliceous coal powder and carbon shale is common and influences surface evolution (Homoki et al. 2000; Sütő 2000, 2007). Losses in the volume of the burnt-out material result in further subsidence and mass movements. The pores are gradually filled up by finer disintegrated components. Physical weathering is obvious on the burnt-out rocks as bare, dark surfaces promote insolation weathering (Erdősi 1987). Under temperate climate, early and late winter freeze–thaw alterations for a few weeks also further mechanical weathering. On disintegrated blocks, physical weathering accelerates and the rate of fractions smaller than 2 mm will decrease within a few months (Sütő 2000, 2007). Spring snowmelt, early summer and autumn rainfalls saturate waste, and induce mass movements on bare surfaces containing coal shale. Excess rainwater enhances gullying and rockfalls on heavily burnt-out slopes of loose material.

Therefore, it cannot be a coincidence that, on the waste tips of opencast mines in three mining districts in East Germany, only in 5 years between 1954 and 1959, a total of 288 mass movement phenomena were recorded, affecting a total of 2.6 billion m^3 of waste (Kézdi 1978).

10.3.2.1 Case Study 1: The Aberfan Slump

On 21st October 1966 in the Welsh town of Aberfan, a waste tip of 110,000 m^3 volume slumped down to bury a school with 116 students and 5 teachers, finally causing a death toll of 144 (Plate 10.3). This disaster is not only an example of environmental hazards, but also has a message for geomorphology. At disposal, no environmental impact assessment was made, and the extracted coal flour waste was

Plate 10.3 The waste tip at Aberfan slumped on 21 October 1966 (Bolt et al. 1975)

piled on the slope with a gradient of 13–14° of Merthyr Hill above the village to a height between 40 and 200 m (Bolt et al. 1975; Bennett and Doyle 1997).

The fine-grained, predominantly clayey-silty waste material with various components was piled on the sandstone with thin, clastic clay intercalations. The slope consisting of impermeable layers was dissected by springs percolating waters and stream valley running towards the town. On the waste tip saturated by abundant precipitation (1,500 mm on annual average) and the abruption of natural runoff, between 1916 and 1966, periodic mass movements were triggered and erosion rills formed. The piling of waste on the two tips was stopped. During the disaster, the lower section of the waste tip liquefied as a result of the high pressure and large amounts of water, lost its stability and moved at an extraordinarily high speed downslope. As a consequence of material deficit, the upper, more stable section of the waste tip was also dislocated, deeply and evenly burying part of the town (Bolt et al. 1975).

10.3.2.2 Case Study 2: The Ádám Valley Waste Tip

The material of this ridge-like tip in Ádám Valley (Borsod Coal-Mining District, Northeast Hungary) with slope angles of 12–20° on a hillside, was piled from coal tubs starting from the valley head, also evidenced by the elongated shape along the former cableway (Plate 10.4). Its southern tongue entirely fills up a minor erosion-derasion tributary valley. Prior to the disposal of the waste, debris cones were formed in the valley by (probably intermittent) watercourses. But the row of waste tips forming an embankment entirely blocks the valley head; therefore surface waters, collecting along the edges of the tips, incised to produce a new erosion

Plate 10.4 Crested waste tips in the Ádám Valley, seen from the main slump (Sütő 2000)

gully, periodically filled up by sediment. In the meantime, the whole deposit was piled along a mine adit and this caused permanent slope instability. The waste tip slumped across the valley side and, on the opposite side, squeezed the unconsolidated molasses on the surface and also triggered mass movements there. Movements could not even be hindered by a natural forest patch (Sütő 2000).

The mass movements on the waste tip in the Ádám Valley are predominantly carpet-like translational slides. On steep slopes between the heap and valley margins, rotational slides occur. In early summer and late autumn, on the 50 m high failure front of the tip, months, 8 m high scarps could be observed (Fig. 10.5). Secondary peak times of scarp retreat result from oversaturation due to spring snowmelt (Homoki et al. 2000; Sütő 2000, 2008). As in natural evolution, regression is most intensive in the central part of the failure scarp; in its foreland typical features, like hollows, spread tongues with crevasses.

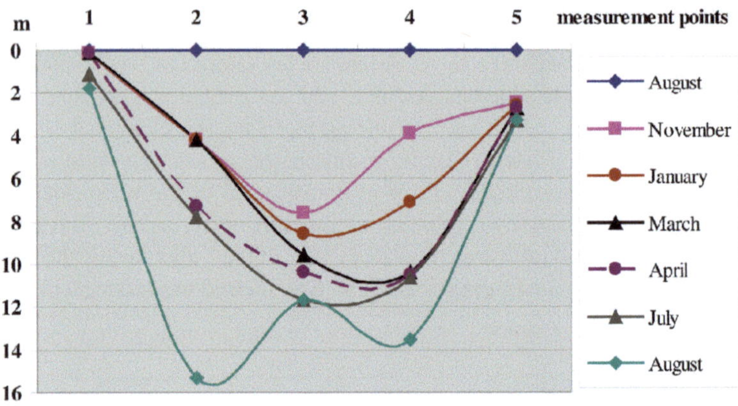

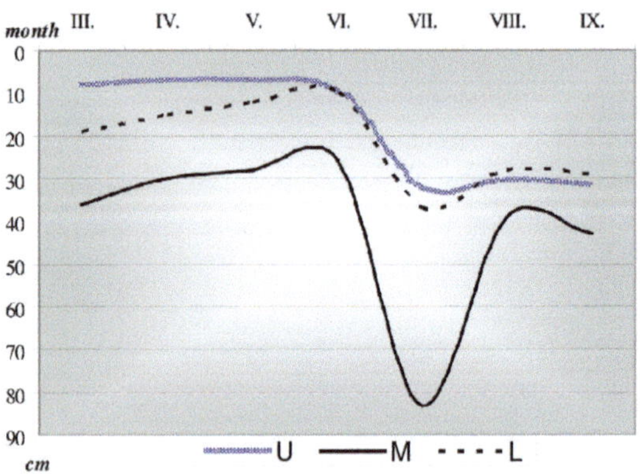

Fig. 10.5 (**a**) Regression of the failure scarp of a waste tip as a function of time (1–5.: number of measurement points on the failure scarp) (Sütő 2008) (**b**) Changes in the depth of erosion rills (at the upper, middle and lower section of the rill) (after Homoki et al. 2000)

On slopes composed of burnt-out blocks, depending on the time of disposal and slope exposure, rows of gullies 0.5–2.5 m deep in various stages of evolution can be distinguished. A minor deepening early spring (10–20 cm) is followed by accumulation of the fine material, which remained after snowmelt (Figure 10.5). As a result of the more intensive early summer showers, even incisions 1 m deep occur. The gullies are then continuously accumulated, with the exception of a minor scour and wash during the November rainfall events. Typical but smaller features of mass transport are also observed in the rills. Incision is more intensive along steeper sections, whereas in the foreland of waste tips alluvial fans are seen. In rills of a later stage of evolution, renewed and seasonal variable erosion can trigger a smaller gully. The seasonal periodicity in incision results in terrace formation (Homoki et al. 2000; Sütő 2000).

On the steepest slopes (35–40°) even the out-burnt waste material cannot be stable and *collapse slopes* with debris fans are formed. Steep failure scarps are also present locally with weathering niches. From overhangs of burnt waste, rock fragments are regularly detached and rockfalls point to slope disequilibria. On the contrary, the role of vegetation is underlined by the fact that no significant erosion began even on steeper slopes if protected by arboreous vegetation. On bare surfaces of waste tips, however, large-scale denudation can be apparent even on more gentle slopes.

10.3.3 Maturity of Waste Tips

Waste tips are also classified based on their *maturity*. Most of the earliest (until the beginning of the 20th century) abandoned small tips in Hungary, associated with small-scale mining, can only be identified by coal shale traces and out-burnt waste powder on the surface.

Waste tips abandoned *before World War II* are more flattened and have gentle slopes. Their material has compacted, their burning has ceased for centuries and humus is forming on their surface and vegetation is spreading. They were afforested the same way as natural slopes. In some parts of the major mining regions, like Upper Silesia, waste tips hide fire-trenches of World War II.

Waste tips from the *post-war* period are the most diversified. In addition to the decaying manual mining, mechanised extraction has become predominant. However, erosion started more slowly on coarse-blocked waste tips due to mechanised cultivation (Gefferth 1986). Although the slopes were stabilised, partly spontaneously and partly by human intervention, both rill erosion and mass movements can be traced. On steeper slopes with no vegetation, erosion processes can renew time after time. As burning only ceases for short spells, topsoil is thin. These tips are less adjusted into the landscape. On waste tips abandoned in the last decades of the 20th century or still cultivated, intensive geomorphic evolution takes place. In this stage, several, already mentioned physical, morphological impact factors have their simultaneous action: physical weathering, chemical weathering, waste burning, gully erosion, deflation, mass movements, etc. The shape of tips could not reach a final, 'mature' stage. Burning and compaction keep up conditions which favour intensive mass movement and erosion processes (Sütő 2007).

10.4 Surface Modification by the Extraction of Fluid Fuels

The *withdrawal of fluids*, as a specific type of fuel, primarily includes hydrocarbons and thermal water to gain geothermal energy. The extraction of drinking, industrial and other waters are discussed in the chapter on water management.

The morphological impacts of drilling are rather limited, but in areas rich in mineral oil, their total surface extent may be a factor in geomorphic evolution and produce *indirect planation landforms*, mostly *negative landforms*, subsidence and depressions resulting from accidents.

When drawing thermal water, in the environs of the deep wells, *depression troughs*, curved cone-shaped surfaces starting from the operational water level form. Such depressions are only permanent features in the case of overproduction. If water table is being constantly lowered, the loading by the weight of the cover sediments can only be counterbalanced by the strength of solid structure and pore water pressure if strength increases due to the lowering water pressure. As a consequence, pore volume decreases and overlying sediments are compacted. The rate of deformation is calculated from Hooke's Law (Somosvári 1989, 2002). As a result of compaction, depressions form on the surface and subsidence will only cease with fluid withdrawal through the rebound of the stationary water level. However, in the case of overproduction, this process becomes irreversible and minor surface deformations will be maintained and a new local base level emerges.

For mineral oil and natural gas withdrawal, the only difference is that the material deficit in the subsurface hollow definitely results in the subsidence of overlying deposits. *Water* withdrawal has minor impacts but permanent depressions are expected in *hydrocarbon* fields (around Lake Maracaibo in Venezuela, in the Niger Delta of Nigeria). The depressions, 1,000–2,000 m across, are usually of oval shape, sometimes merging into more extensive depression fields.

One of the best-known examples is the Long Beach-Wilmington area in California. The vast oil field under the city was discovered in the 1920s, subsiding areas were first identified in the 1940s, when wharves were flooded and pipelines, buildings and railways cracked (Fig. 10.6). Since the 1950s, vertical displacements reached the order of metres, thus in the subsurface strata horizontal shearing occurred. Furthermore, sudden movements also triggered small-scale earthquakes at shallow depths.

By the 1960s, the value of the maximum subsidence reached 10 m, and therefore, remedial actions were required. Water was injected into unused fields to increase pore pressure. The experiments to arrest subsidence proved to be successful; the initial elevation of the terrain, however, could not be restored; surface rebound remained insignificant; most of the harbour is still continuously flooded. Large-scale displacements have not occurred in the city since then (Bolt et al. 1975).

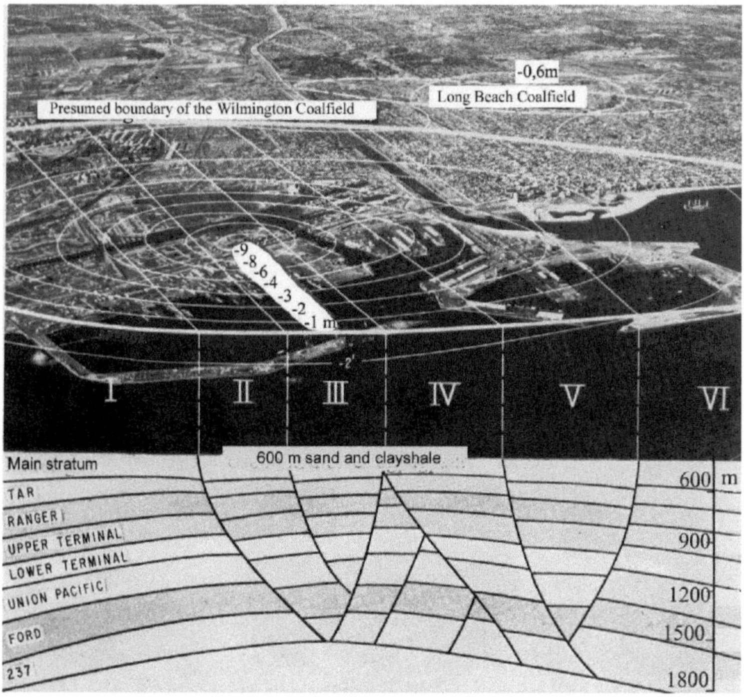

Fig. 10.6 A model of subsidence at Long Beach, California (after Bolt et al. 1975)

In Hungary, surface movements were reported in relation to hydrocarbon extraction in the Southern Great Plain at Pusztaföldvár, Medgyesegyháza and other sites, while the withdrawal of drinking water had such consequences in the environs of Debrecen. Between 1927 and 1966, land subsidence took place in the 10–15 km environs of Debrecen. A depression of maximum 40 m depth was produced under the centre of the town (Bendefy 1968; Orlóczi 1968). Its impacts are discussed in the chapter on water management.

Surface alterations are also caused by accidents, e.g. the *explosion of hydrocarbon wells*. A special feature is that they are endogenic in origin, similar to volcanic activity. An example is the *crater* at the village of Nagyhegyes, formed on 24th August 1961 (Fig. 10.7). The explosion took place as a result of excess pressure of the gas emitted during drilling. The gas eruption was followed by gas burning, then the rupture and dispersal of surface rocks. Of the 172,000 m^3 loess and silty alluvium, a 4–6 m high regular circular rampart was built. Two days later, the eruption was choked by water accessing the system and, as with post-volcanic activities, further operated as mud volcano for 3 years when its activity ceased for good. A crater pond of nearly 8,000 m^2 area and 931 cm depth formed from ground and rainwater. It is shrinking in size as it is filling up. These sizes indicate a slow decrease as a consequence of accumulation. Regular rotational slides shape the edge of the pond basin (Borsy 1967).

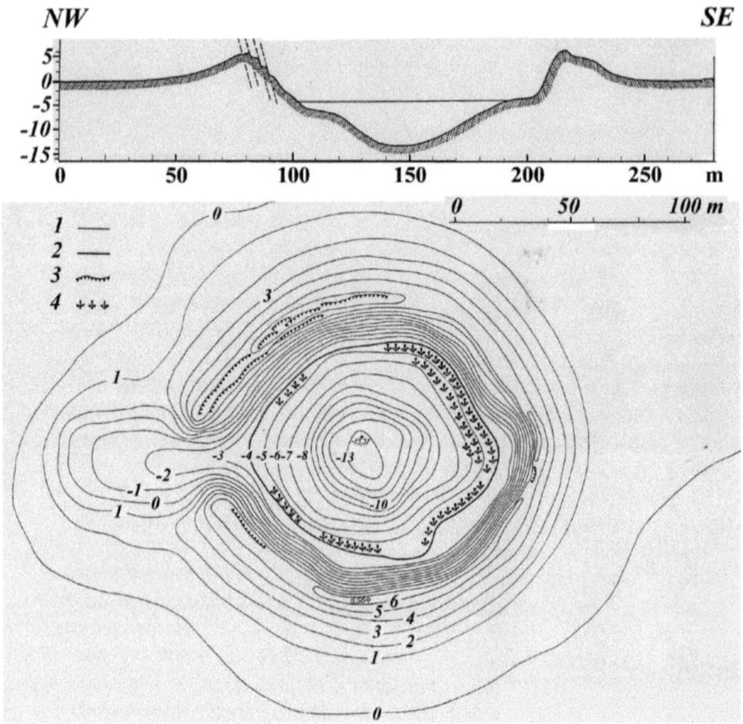

Fig. 10.7 The crater at Nagyhegyes in profile and plan view 1. contour line; 2. edge of pond; 3. landslides; 4. reed-bed (after Borsy 1967)

10.5 The 'Geomorphological Cycle' of Mining Landscapes

Although they are only the by-products of excavation, mining landforms (probably) have the most remarkable impact among human activities. They modify geomorphic evolution to reach a new state of equilibrium. As whole regions are affected, researchers mention *'mining landscapes'* (Nir 1981; Szabó 1993; Sütő et al. 2004) and distinguish the periods of intensive human intervention as anthropogenic landscape cycles (Wach and Szczypek 1996).

Mining plays a significant part in the re-activation of movements at molasse areas susceptible to surface movements in areas like the Borsod Coal-Mining Area, the Mecsek Mountains or Upper Silesia; explicitly montanogenic and natural landforms co-exist. The rate of surface evolution is necessarily influenced by mining on undulating valley sides, recent erosion rills and on minor slumps (Erdősi 1987; Homoki et al. 2000).

Landscaping strives to achieve a surface pattern favourable from the ecological point of view. Information on waste tip 'maturation' and subsidence processes are needed to control geomorphic evolution and to design efficient landscape management.

10.6 Case Study: Uranium ore Mining in the Mecsek Mountains

Dénes Lóczy

In strict secrecy and under complete control from the Soviet Union, uranium ore mining started in Hungary, west of the city of Pécs, in 1955. Until 1964 the extracted raw ore was transported to the Soviet Union (to Estonia) and then processed in place and concentrated ore was exported. Mining and processing, already unprofitable and without strategic significance, ceased in 1997 and was followed by a survey of environmental damage and a plan for land reclamation (Lendvai-Koleszár et al. 2003).

The total mining area, comprising six plots reserved for mining in 1967 and the ore beneficiation plant, was 65 km^2. The waste tips, percolation prisms (for gaining uranium ore from percolating waters), sludge reservoirs, ventilation shafts, drifts, ore passes and other facilities extended over the area of five more villages beyond the central settlement of Kővágószőllős. As many as 362 buildings, including store spaces for explosives and social establishments, were raised in this area.

The large-scale topographic impact of uranium mining derives from the low ore concentration and, consequently, from the huge amounts of refuse rock that has to be moved. Over the four decades of mining, 18,000,000 m^3 of rock, mostly Upper Permian greyish sandstones were recovered, and that equals the total volume of underground cavities (240,000 m^3 of vertical shafts, 8,250,000 m^3 of horizontal galleries and 9,345,000 m^3 of workings and other cavities). A total of ca. 46,000,000 tonnes of rock, including 25,400,000 tonnes of raw ore (about one half of the total reserves), was brought to the surface. As the average metal concentration was 810 g/t, the mine supplied 20,800 tonnes of uranium metal.

In this area, quarrying predates ore mining: the hanging wall of ore-bearing rocks, Permian red sandstone, was used as a building material. In the second half of the 20th century, next to old quarry pits, waste was accumulated partly on Permian sandstone surfaces, partly on Upper Pannonian (Pliocene) gravelly sands. The tips did not only result in the aesthetic deterioration of the landscape but destroyed the fertile forest, meadow and alluvial soils, upon which refuse was dumped. In addition to polluting waters with uranium, radium and various organic compounds, uranium mining also resulted in soil contamination with oil, acids, manganese ore and carbonates.

As a consequence of the complex tectonic conditions of the Mecsek Mountains (Chikán and Konrád 1982) and the extraction of subsurface waters, the technological practices associated with mining (explosions, headwall working) had an almost immediate impact on the surface. Gradual subsidence and abrupt collapses were both experienced in the area of Shaft 1. The buildings which belonged to Shaft III as well as residential homes in the villages of Kővágószőllős and Cserkút were damaged. Even above the much deeper shafts IV and V, traces of subsidence were observed. During the stabilisation of underground workings special care had to be applied: shafts opening onto the surface had to be backfilled and adjusted to the rate of the compaction recorded; repeated backfilling was necessary. As a matter of course, all substances potentially dangerous to the environment had had to be removed before.

The water level depression funnel of ca. 42 km^2 area, elongated in east to west direction and established to make mining possible, had both regional and point-like impacts. With reducing pore pressure, the hanging wall subsided as a single mass, more or less uniformly, in the vicinity of the deeper shafts. At the same time, water levels in the drilled and dug wells dropped dramatically and some of them even became dry. A severe consequence of water table lowering was the change in groundwater flow. The recharge of the two major confined groundwater aquifers of Pannonian age (Pellérd and Tortyogó), from the direction of the Mecsek Mountains, which are crucial to the water supply of the city of Pécs, remarkably reduced in volume. In order to protect water reserves and to prevent percolating waters from reaching them, 5,000,000–6,000,000 tonnes of material had to be relocated from the percolation prisms to the tailing tips situated to the west.

The tasks of land reclamation after mining fell into two groups:

- the abandoned *underground* cavities generally had to be collapsed (after recovering machines and other objects from them) very cautiously to prevent the release of radon gas and the accumulation of water in them;
- *surface* structures had to be demolished, contaminated areas cleaned, wastes disposed and man-made landforms secured to present no hazard to the environment.

A major task was to 'obliterate' local percolation prisms of 17 m height or, if it was not possible, to make them look more natural. This was done by landscaping and a large-scale reduction of slope angles; the measures led to successfully restoring the ca. 1° southern slope of the pre-mining foothill surface of the Mecsek Mountains.

To inhibit radon emission, sludge reservoirs received a 1.5 m deep earth cover (30 cm of clay, 30 cm of compacted loess, 30 cm of sand and another 60 cm of uncompacted loess). The nine waste tips of ca. 90 ha total area and ca. 10,000,000 m^3 total volume) were mantled by 1 m thick compacted sediment. The total material demand of land reclamation here amounted to ca. 2,500,000 m^3 of Pannonian sediment. Before spreading the sediment cover, the unstable sludge with an oversaturated core in the reservoirs had to be stabilised by desiccation and applying geonets and geotextiles on their surfaces. On sloping surfaces, the sediment cover is endangered by erosion after intense showers. To prevent erosion of various types (sheet wash, rill and gully erosion as well as piping), quickly growing arboreous vegetation had to be planted. Trees with high transpiration also promote the dewatering of accumulated waste. In view of radiation and the oxidation of sulphides, 'recultivation' proper, i.e. an agricultural use of percolation prisms and sludge reservoirs is not feasible for several decades (for a minimum period of 30 years). Therefore, the continuous monitoring of environmental conditions is indispensable, and a system for this monitoring has been designed and put into operation. In addition, the design of topography has to meet the requirements of surface drainage (through establishing trench drains).

Based on radiation threshold values the area has been divided into zones of unlimited and those of limited utilisation. Most of the latter surfaces were grassed or planted by arboreous vegetation in the first stage of reclamation.

References

Ádám L, Pécsi M (eds.) (1985) Mérnökgeomorfológiai térképezés (Engineering geomorphologic mapping). MTA FKI, Budapest

Bendefy L (1968) Debrecen városi belsősége süllyedésének hidrogeológiai vonatkozásai (Hydrogeologic Aspects of Settlement Subsidences Observed in Debrecen). Hidrológiai Közlöny 48 (12): 549–559.

Benke I, Reményi V (eds.). (1996) A magyar bányászat évezredes története 1-2. (A millenium history of mining in Hungary Vol. 1–2). Országos Magyar Bányászati és Kohászati Egyesület, Budapest

Bennett MR, Doyle P (1997) Environmental Geology: Geology and the Human Environment. John Wiley and Sons, Chichester

Bertalanfy B, Derecskei Gy, Jakucs S, Kaminszky L, Kemény Gy, Kreffly I, Lostorfer R, Papp B, Rem L, Remény G, Simon L, Sivák I, Szalai L, Szepessy A, Székely T, Sztermen G, Tóth P, Váci L, Zsámboki L (1986) 200 éves a borsodi szénbányászat (200 years of coal mining in Borsod). Borsodi Szénbányászati Tröszt, Miskolc

Bolt BA, Horn WL, Macdonald GA, Scott FR (1975) Geological hazards. Springer-Verlag, Berlin: Heidelberg: New York

Borsy Z (1967) A Nagyhegyesi-kráter (The crater at Nagyhegyes). Acta Geographica Debrecina, 12–13 (1966/67): 84–100.

Chikán G, Konrád Gy (1982) A nyugat-mecseki földtani térképezés újabb eredményei (Current results of geological mapping in the Western Mecsek Mountains). M Áll Földt Int Évi Jel az 1980. évről

Dávid L (1999) Antropogén hatások a tájban a Mátra déli területein (Anthropogenic impact on the landscape in the Southern Mátra Mountains). In: Füleky Gy. (ed.): 2. Tájtörténeti Tudományos Konferencia: A táj változásai a Kárpát-medencében/. – Konferenciakiadvány. – Gödöllő

Egerer F, Namesánszky K (1990) Magyarország meddőhányóinak katasztere (A cadaster of the waste tips in Hungary). Miskolci Egyetem, Miskolc

Erdősi F (1987) A társadalom hatása a felszínre, a vizekre és az éghajlatra a Mecsek tágabb környezetében (The impact of society on the surface, water and climate in the broader environment of the Mecsek Mountains). Akadémiai Kiadó, Budapest

Fodor T, Kleb B (1986) Magyarország mérnökgeológiai áttekintése (An engineering geologic review of Hungary). MÁFI, Budapest

Gefferth K (1986) A meddőhányók és bányatavak hasznosításának lehetőségei és gondjai (Possibilities and difficulties of the use of waste tips and mine ponds). Bány Koh Lapok 119 (3): 181–186

Homoki E, Juhász Cs, Baros Z, Sütő L (2000) Antropogenic geomorphological research on waste heaps in the East-Borsod coal basin (NE Hungary). In: Z Badań nad wpływen antropopresji na środowisko/Rzętały, M. (red.). – Sosnowiecz: Studenckie Koło Naukowe Geografów UŚ

Hoványi L, Kolozsvári G. (1989) Geodézia és Bányaméréstan: Bányaméréstan II. (Geodesy and mining metrology. Mining metrology Vol. 2). Tankönyvkiadó, Budapest

Jambrik R (1993) Külfejtések víztelenítésének hatására jelentkező felszínmozgások meghatározása (Defining surface movements apparent). Miskolci Egyetem, Miskolc

Kerényi A (1995) Általános környezetvédelem (Environmental protection). Mozaik Oktatási Stúdió, Szeged

Kézdi Á (1978) Talajmechanika: Példák és esettanulmányok (Soil mechanics – Precedents and case studies). Tankönyvkiadó, Budapest

Kleb B (1990) Az ásványi nyersanyagok bányászatának, felhasználásának környezeti hatásvizsgálata távérzékelési módszerekkel (An environmental impact assessment of the mining and use of fossil raw materials by applying remote sensing methods). Földt Kut 33 (3): 41–49

Ládai JT (2002) A külszíni süllyedés időbeni lefolyásának vizsgálata (A study of temporal course of outcast subsidence). Bány Koh Lapok 137

Lendvai-Koleszár Zs, Hideg J, Csáki F (eds.) (2003) Szilárd ásványbányászati alprogram. Uránbányászat (Solid mineral mining sub-programme. The mining of uranium). Környezetvédelmi és Vízügyi Minisztérium, Budapest. (Kármentesítési Füzetek 9)

Louis H, Fischer K (1979) Allgemeine Geomorphologie. Berlin : New York

Martos F (1955) Újabb adatok a külszíni alakváltozások számításához (Current data to the calculation of outcast shape alterations). Bány Lapok 10: 538–544

Martos F (1956) A külszíni süllyedés számításának egy közelítő módszere (An approximation to the estimation of outcast subsidence). Bány Kut Int Közl 1: 3–12

Martos F (1958) A külszín elmozdulását befolyásoló tényezők (Factors influencing the surface displacement). Bány Lapok 91 (1): 367–372

Nir D (1981): Man, a geomorphological agent. Keter Publishing House, Israel; D. Reidel Publishing Co., Holland

Orlóczi I (1968) A felszín alatti vízkészlet igénybevételének vizsgálata Debrecenben a terepszint süllyedésének mérése alapján (Investigation of the Water Balance Based Upon Measurements of Land Subsidence Caused by Ground Water Withdrawal in Debrecen). Hidrológiai Közlöny 48 (5): 205–213.

Pécsi M (1991) Geomorfológia és domborzatminősítés (Geomorphology and relief classification). MTA FKI, Budapest

Sherlock RL (1922) Man as a geological agent. London.

Somosvári Zs (1989) Geomechanika II. Tankönyvkiadó, Budapest

Somosvári Zs (1990) Geomechanika I. Tankönyvkiadó, Budapest

Somosvári Zs (2002) Felszínemelkedés jelenségek okai (Reasons for land upheaval phenomena). Miskolci Egyetem Közleménye: Bányászat 62: 35–46

Sütő L (2000) Mining agency in the east Borsod Basin, North-East Hungary. In: Jankowski AT, Pirozhnik II (ed.): Nature use in the different conditions of human impact. Minsk; Sosnowiec: Studenckie Koło Naukowe Geografów UŚ

Sütő L. (2007) A szénbányászat geomorfológiára és területhasználatra gyakorolt hatásainak vizsgálata a Kelet-borsodi-szénmedencében (Research of the effects of coal mining on the geomorphology and the land-use system in the East Borsod Coal Basin). PhD dissertation, DE Földtudományi TCs, Debrecen

Sütő L. 2008. Aktív tömegmozgásos folyamatok kőszénbányászati meddőhányón (Active mass movements processes on waste tips of coal mines). In: Kiss T. – Mezősi G. (szerk.) Recens geomorfológiai folyamatok sebessége Magyarországon. Földrajzi Tanulmányok Vol. 2. – Szeged, SZEK Juhász Gyula Felsőoktatási Kiadó. pp. 143–149.

Sütő L, Homoki E, Szabó J (2004) The role of the impacts of coal mining in the geomorphological evolution of the catchment area of the mine at Lyukóbánya (NE-Hungary). In: Lóki J, Szabó J (eds.): Anthropogenic aspects of landscape transformations III. Hungarian – Polish Symposium, Debrecen 28–29. September 2004

Szabó J (1993) Antropogén geomorfológia (Anthropogenic geomorphology). In: Borsy Z (ed.): Általános természetföldrajz. Nemzeti Tankönyvkiadó, Budapest

Wach J, Szczypek T (1996) Преобразования рельефа местнсти в районах горнодобыающй проышенности вследтвие оседаний грунта (на примере катовицкого воевдства). In: Географические проблемы природопользования b условиях антропогенной деятельности. Минск : Белорусский Государственный Университет и Белорусс кое Географическое Общество.

Chapter 11
Water Management

József Szabó

Abstract Water management (river regulation and flood control) is one of the earliest human interventions into geomorphic evolution. River regulations and flood control are represented by two basic structures: dams as positive and channels as negative landforms compared to the adjacent surfaces. Engineering structures of water management do not only alter total amount of water available but also its spatial and diurnal distribution. Locally increasing or decreasing water discharge results in changes in the parameters of sediment transport and influence geomorphic evolution. Consequently, new landforms develop locally and others begin to decay. Water management measures, however, can often significantly affect sediment transport not only locally but on long river sections. During hydraulic engineering works, both increase and decrease shear stress may occur, hence mass movements can be initiated. Underground water management is a common cause of non-slope mass movements (ground subsidence). Their relevance has become well acknowledged in the last century due to the rapidly increasing use of underground waters. Alterations of coasts and lake-shores have been made for basically two reasons: either for the purpose of coastal defence ('passive' intervention) or for a particular economic activity ('active' intervention). Water management covers both kinds of activities.

Keywords Water management · River regulation · Flood control · Land reclamation

11.1 Introduction

It is pointless to argue whether agriculture or water management (*river regulation* and *flood control*) was the first human intervention in geomorphic evolution. The first civilisations in history could only be established and sustained through an intensive use of water; their share based on the surplus production advanced their

J. Szabó (✉)
Department of Physical Geography and Geoinformatics, University of Debrecen, 4010 Debrecen
e-mail: wagner@puma.unideb.hu

J. Szabó et al. (eds.), *Anthropogenic Geomorphology*,
DOI 10.1007/978-90-481-3058-0_11, © Springer Science+Business Media B.V. 2010

rapid development (and greater power) compared to their neighbours. Potamic cultures in arid–semiarid regions of the subtropical zone, including the Nile Valley, Mesopotamia, the Indus and Ganges regions and the giant rivers in China were all based on intensive use of water. This water use meant, from the beginning, more than just exploiting the potentials offered by natural water flow. According to radiocarbon dating, in the delta of the River Nile, land cultivation carried out for 7,000 years utilised not only flood water as claimed by, e.g. Legget (1962); to the south of Memphis, as early as more than 6,000 years ago, a rock dam 457 m in length and exceeding 15 m in height was built and operated for 4,500 years. A similar dam from the Prehistoric times at Marouk on the River Tigris was functional until the 13th century. A dam in Yemen, 3,700 years in age, 3.2 km in length, 36.5 m in height and 152 m in width was breached during a flood in 300 A.D. For the construction of the famous dam of Saad el Kafara near Heluan (Egypt), during the period between 2950 and 2750 B.C., more than 100,000 tonnes of building material were used. A long list of early water management structures directly altering the surface, among them, the impressive ancient canals, can be mentioned. However, their geomorphic impacts during their long time of operation has more geomorphological significance than their spectacular remains in the landscape.

The same applies to another major field of water management and engineering, i.e. *coastal engineering* and *protection*. When sometime later, the marine cultures of the East Mediterranean (Phoenician, Aegean then Greek) flourished, the fundamental interests of shipping demanded local transformations of coasts. It was first represented by the construction of harbours, logically followed by their ramparts. In many cases, harbour facilities themselves also altered the tendency of coastal development, even if limited to short sections but often in unfavourable direction. Therefore, extensive interventions were required both along the coasts and in their foregrounds. Major interventions usually triggered further impacts; consequently, certain coastal sections were fundamentally transformed. This made it impossible to operate the harbour sometimes, which then had to be abandoned. These coastal processes reoccurred many times and at many locations on Earth later in history, due to larger-scale human interventions.

As river management and flood control became more widespread activities in time and their (morphological) impacts also extended, in some cases, mostly at river mouths, both activities had a cumulative impact. Thus, the direct and indirect social impacts on fluvial and coastal processes intensified.

A study into the changes resultant from the use of *underground water* also belongs to the impacts of water management on the surface (Szabó 1993). However, it should be kept in mind that in this sense, water management measures here, in many cases, are taken in accordance with demands by other sectors of economy, even indirectly. Water exploitation can fulfil, e.g. either industrial agricultural or even urbanisation needs. The extraction of underground water reserves to be used as raw material, however, is part of mining. Therefore, a rather high number of segments of social activities can be blamed for the often negative – not only geomorphological – consequences of the intensively growing use of underground water. The above cannot only be claimed about the utilisation of underground water.

As most water management interventions serve other social activities, many of them are related to the impacts of other branches of the economy, their rigid distinction is not justified.

As already seen from the above, when thinking about the geomorphological impacts of water management, consequences of (expansive) activities aimed to secure or even improve social welfare and of the protection measures against damage by water (e.g. the already mentioned coastal defence, or flood control) should equally be taken into account. This is especially the case for the most classic method of flood control, i.e. the construction of embankments, an explicitly geomorphic activity. Water management was one of the most required and inevitable activities of society. It is also the case today and even more so, due to the increasing awareness of the value of water. The account of geomorphological consequences of water-related works is also supported by the accumulating evidence indicating that precedent interventions aiming at both exploitation and control relatively often resulted in negative (geomorphological) consequences, too. In order to avoid these consequences, planning should to be carried out with better expertise and care in the future.

11.2 Geomorphologic Impacts of River Regulations and Flood Control

11.2.1 Relevant Landforms

From the geomorphologic point of view, the two basic structures of such works are *dams* as *positive* and *channels* as *negative* landforms as they rise above or deepen below their environment, respectively. Dams facilitating the elevation and regulation of water level are mostly *transversal*, and almost exclusively also *primary* landforms. Non-desirable inundations are mostly intended to be avoided by *longitudinal* (flood-control) dykes. Apparently, they are also *primary* landforms. Redirection of water can be achieved by also *primary* but *negative* landforms (canals). Both activities lead to the creation of *secondary* landforms usually in the opposite way. Material for positive landforms must be excavated and this is usually associated with creating a depression. This is especially the case in lowlands. The material for the flood-control dykes is mainly obtained from areas next to the embankment, therefore embankments are usually accompanied by a low waterlogged strip, a row of navvy pits on the floodplains. Especially in densely populated areas under intensive land use, such secondary negative landforms today are patches representing an increasing ecological value, which over time will turn into quasi-natural in character. Also, secondary landforms are basically embankments of the excavated material piled up on both sides during the deepening of canals, which can also facilitate water management; therefore, they have primary connections. In case, in the environs of dyke constructions, natural or anthropogenic positive landforms can be found, their partial or entire removal means the *levelling* of the surface. In the history of river regulation and flood control in Hungary, a great amount of material of wind-blown sand dunes or tumuli was incorporated into the body of dykes. In mountainous regions, the extraction of hard rocks for the purpose of water management also leads to levelling. The transportation of rock masses, often over long distances, and their accumulation along lower river sections, can even change the lithological character of the surface locally and in the long term.

The direct geomorphological significance of *transversal barrages* to water-courses is represented by the amount of material moved to create such landforms. The 100,000 t mass of the already mentioned Dam of Heluan is dwarfed by those of the giant barrage dams of the 20th century. The Grand Coulee Dam on the Columbia River – as one of the early examples of the giant dams built after World War II – is 20 million tonnes (equalling to approximately 4 times the weight of the Cheops pyramid). In the past 50 years, dams manifold greater were also erected (on the River Vahsh of Tajikistan, two dams higher than 300 m are double the size of the Grand Coulee). The increase of the geomorphological significance of dams is well illustrated by the fact that the number of dams higher than 15 m on the Earth, between 1950 and 2000 increased from 5,750 to more than 41,000, 45% of which are found in China, 38 on the Yangtze (Changjiang) River alone – and 349 barrages with a height exceeding 60 m were under construction in a single year (1998) (Der Fischer Weltalmanach 2002). Barrages have been constructed for 14% of all watercourses worldwide, and due to their constructions, 40–80 million inhabitants had to be relocated. In this sense, Kollmorgen's claim in 1953, on the 'Floodplain cannibalism' due to barrages is well justified (Kollmorgen 1953). (A good example is the Grand River, a tributary of the Missouri where nearly 61,000 ha of valley floor had to be flooded in order to protect 86,000 ha of land, and the relocation of 5 towns and several hundreds of villages was also necessary. Further indirect cost was the removal of most of the fine alluvial sediment from the lower valley floor areas.) In 2000, The World Commission on Dams, established in 1997, concluded that barrages despite their great contribution to human development have, in many cases, significant negative impacts on both society and the environment. This might question the necessity for many barrages, and more circumspection is needed during their construction and use in the future. Such questions are also raised from the geomorphological point of view.

Although the cross-sectional dimensions of *flood-control dykes* are lesser than those of barrages (along the rivers in Hungary, they rarely reach the height of 10 m; however, the height exceeds 15 m at St. Louis along the Mississippi), their construction demanded the removal of more material due to their length. The length of dykes in Hungary exceeds 4,300 km (Fig. 11.1); in the Netherlands, 3,300 km of dams were constructed, whereas more than 4,000 km were constructed along the Mississippi. Their geomorphologic relevance is highlighted by the fact that such dykes are, in general, *the most outstanding positive landforms in their environment.*

Today, experts in hydrology (as well as the general public) are concerned worldwide about the fact that flood-control dykes cannot provide adequate protection at many locations. The increase of peak flood waves is a phenomenon known not only

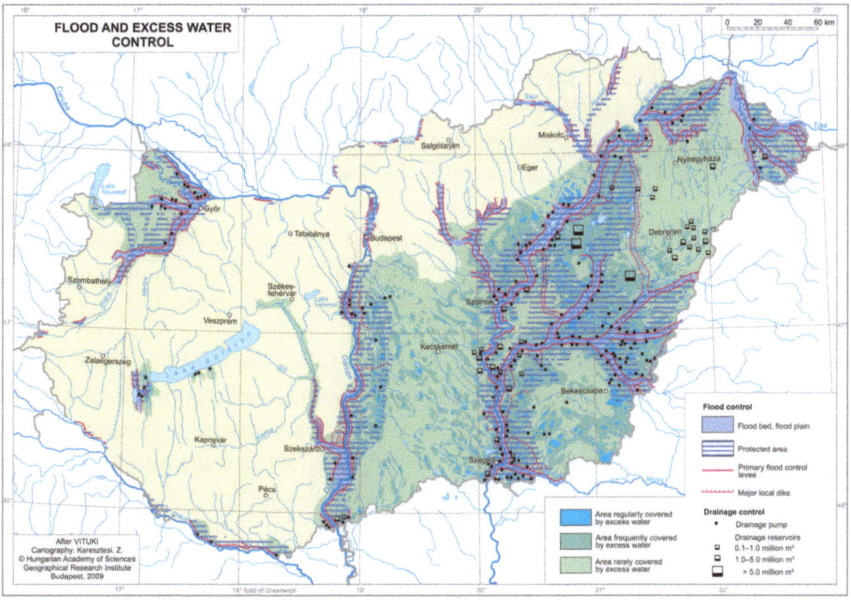

Fig. 11.1 Construction of flood-control dykes in Hungary (VITUKI Environmental Protection and Water Management Research Institute, after Alföldi (2000))

from the rivers of the Carpathian Basin. This trend required the increase of dyke size in the central part of the basin (in Hungary, too) in the past century. The reasons for this are manifold and often controversial; however, the most obvious solution was to raise dykes. Recently, in many countries, doubts arose whether this is really the best way of flood control. In case new solutions are found (see e.g. the plans for the construction of 10–14 emergency reservoirs proposed in the Vásárhelyi Plan), it can be claimed that the geomorphological and visual significance of flood-control dykes will grow worldwide in the near future.

For both dam types (transverse and longitudinal) it can be true that various impacts of old and usually smaller dams manifest after a long period of time. Large-scale dams of the modern age had less time to reveal their impact on the environment, which is likely to show in the future. They definitely have to be taken into account.

The most relevant group of *negative landforms* includes *canals*. At first, they were mostly dug in order to conduct river water to cultivated lands (in Egypt and Mesopotamia); however, the construction of a drainage ditch network facilitating the reclamation of waterlogged marshlands was also initiated (around Thebe in Greece, draining of the Pontini Marshes or the Maremmas near Rome). The largest canals were built for transportation purposes (the Ptolemaios Canal in Egypt or the Great or Imperial Canal of China between Beijing and Hangzhou used for 4,000 years as well as medieval canals in Europe also advancing transportation). A rather unique type was represented by aqueducts used mainly in the Roman Empire for communal water supply. Although such structures cannot really be regarded as a direct subject of anthropogenic geomorphology, they are worth the attention due to their environmental, and among others, morphological impacts.

Irrigation societies established and maintained the canal network. Either increasing or decreasing water discharge changes the geomorphic impact of natural watercourses. This is observed, e.g. today in Central Asia as a result of the irrigation of the water of rivers feeding Lake Aral. It has an indirect but rather significant morphologic impact in the rapid contraction and division of Lake Aral into partial lakes as well as the creation of several thousand square kilometres of new mainland. Similar processes can be observed around the Caspian Sea, along the River Jordan and at many other locations.

Canals established related to river regulation and flood control are predominant anthropogenic landforms of broad valley floors and lowlands in the past centuries. Their construction was basically initiated by a principle applied for flood prevention, claiming that flood peaks and their duration can be reduced mainly by increasing river gradient. To achieve this, the most obvious solution is to reduce river length. This method can be effective mainly for meandering rivers, cutting through bends. Cut-offs result in new *channels* and *ox-bow lakes*. Ox-bows are basically an indirect result of human activity (the landform itself is not created by humans), and cannot be regarded as a primary feature as it was not deliberately produced. Channels cutting through meander necks can only be partially regarded as the direct and exclusive products of human activity. The new river bed at such cut-offs usually developed along 'lead ditches' of significantly narrower cross-section. Such ditches were broadened into a main channel by the river itself. This was so typical for the regulation of the Tisza River in Hungary (for details, 11.2.1) where some of the cut-offs were not adjusted by the river but decayed or had to be extended. Examples are found in higher numbers mainly along the Lower Tisza.

The geomorphologic significance of ox-bow lakes formed by cut-off is well indicated by data related to the flood-control construction of some known watercourses. In Germany, the regulation of the Upper-Rhine section with especially high flood risk between Basel and Bingen was initiated, in accordance with the proposals by J. G. Tulla, in 1817–1818 by cutting-off eight curves that was followed by further cross-cuttings until 1880 by which the river was shortened here from 354 to 237 km (Mock 1992). Among the similar but relatively more recent works, 16 vast cross-cuttings can be mentioned by which a 600 km section of the Mississippi between Memphis and Baton Rouge was shortened by 270 km.

For the construction of canals for various purposes vast amounts of material had to be moved and major landforms created. To cite only one example, the 202 million m^3 of earth removed during the construction of the Moscow–Volga Canal during the early period of the great nature transformations in the Soviet Union, in 1937, approximately 80-fold surpassed the volume of the Cheops Pyramid. It was followed by the even larger-scale canal constructions of the 20th century. Most recently, a 360 km long canal section is being built along the Sudan section of the White Nile in order to avoid the Sudd marshlands within the framework of the Jonglei Canal Project that carries an annual surplus water of nearly 5 km^3 to North

Sudan and Egypt. (Obviously, this amount of water will be a deficit for Southern Sudan and further deteriorate the conditions of local residents.) Although such large-scale works of water diversion as above will raise justifiable questions on the further cultivation of extensive areas and the future of their environment, they are certainly to be regarded geomorphological features.

The landform types analysed do not include the entire range of features created during channelisation as man-made landforms. Other engineering interventions in the river channel or on its margin (groynes, steps, revetments, etc.) have not yet been mentioned. Their sizes are usually far behind those of landforms discussed above. On the other hand, they are mostly found on the river bed and, consequently, neither their morphological appearance nor their impacts can be compared to that of facilities of high-water regulation, which occasionally transform the face of whole regions.

11.2.1.1 Case Study 1: Geomorphologic Impacts of the Regulation of the Tisza River

The Tisza River has the largest catchment in the Carpathian Basin (157,000 km^2). The trunk river of the system had a length of 1,419 km, from its source in the NE Carpathians to its confluence with the Danube River, when river channelisation started in the early 19th century. A total length of 1,213 km fell onto the plains of the basin. From the Ukrainian–Hungarian border to the confluence, the river was of meandering mechanism, and its major tributaries in the plains were also meandering. Its gradient at the 955 km long Hungarian section was remarkably low, even in comparison with other rivers in the plains, only 3.1 cm/km; but in many locations not exceeding even 2 cm. Thus, large areas were affected by its floods (in total, about 26,000 km^2), and remained waterlogged for months. In most of the Great Hungarian Plain, no intensive farming could be carried out and, when demanded by the emerging capitalist development, flood-control works were launched. The regulation initiated by István Széchenyi (1791–1860), was basically implemented in accordance with the conception and plans of the engineer Pál Vásárhelyi (1795–1845), from 1846 to practically the end of the 19th century. His main concept was to increase the gradient of the river by cutting-off curves (thus by reducing its length), to accelerate its flow and thus, to reduce the duration of floods. To decrease flooded areas, embankments were constructed – until the mid-20th century, about 4,500 km long in the Tisza River catchment – Lászlóffy 1982); consequently, floodings were restricted to a floodplain of 1,500 km^2. By cutting-off 114 curves of the River Tisza, its total length decreased to 955 km, and its gradient was nearly doubled (increased to 5 cm/km). Similar works of even larger scale were carried out on its tributaries, e.g. on the Körös and Berettyó rivers, where 265 curves were cut through and the rivers shortened to only 37% of their initial length. Accelerated flow significantly reduced the duration of flood waves, rivers had been cut and the low water levels dropped on several sections of the Tisza River even by 3 m. All this was associated, on most of the protected floodplain, with dropping groundwater table and the alteration of soils (at many places through alkalisation), and the lack of regular yearly

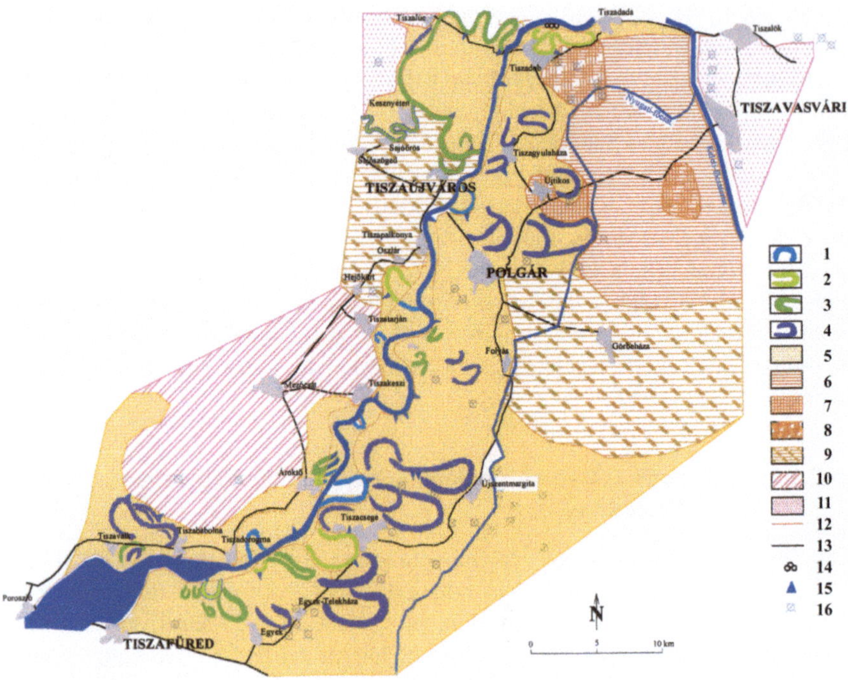

Fig. 11.2 Relict meanders of various age and type in the Middle-Tisza Region in Hungary 1 = channel cut within the dyke, 2 = channel cut outside the dyke, 3 = naturally cut-off meander outside the dyke, 4 = older cut-off meander well recognisable on topographical maps, 5 = young levèe with a relatively uninterrupted system of channels and levèes, 6 = levèe with still recognisable channel pattern, 7 = old levèe with a vague pattern of channels, 8 = aeolian sand surface, 9 = fluvial pattern recognisable only on satellite images, 10 = terrains mainly formed by tributaries, 11 = high terrains not affected by floods, 12 = embankment, 13 = paved road, 14 = backswamp in development, 15 = canal, 16 = tumuli (Tóth et al. 2001)

inundation resulted in relative desiccation. On the contrary, significant accumulation began in the active floodplain, the exact rates of which are currently being studied. On the protected floodplain converted to a cultivated grassland within one and a half decade, fluvial geomorphic action ceased, and in the recent morphology of the Great Plain along the Tisza River the most remarkable landforms as well as the ecologically most valuable patches are the traces of the several hundred ox-bow lakes, naturally or artificially formed (Fig. 11.2).

11.2.2 Changes in Fluvial Action

Engineering structures of water management do not only alter the total amount of water available but also its spatial and diurnal distribution. Locally increasing or decreasing water discharge will result in changes in the parameters of sediment transport and influence geomorphic evolution. Consequently, new landforms develop locally and others begin to decay. On the whole, water management will have a primary impact on the material removal (erosion/accumulation) relevant to geomorphology. Such changes will often become apparent within a rather short period of time. Under the mouth of the diversion canal at Gabčikovo, significant alterations in the bed of the Danube River were observed a few years later.

Large-scale *dams* have opposite headwater and tailwater impacts (Fig. 11.3). Decelerated water flow will result in upstream deposition. If impoundment results in river widening, the alluvium of the delta will fill up the bottom relatively evenly both longitudinally and in cross-section. At the upper end of the transgressing delta, the alluvium accumulating to the water level will build a fan. In the impounded reservoir sediments – in the case of larger lakes, even suspended load – deposit and thus downstream of the dam, the river will transport virtually no load (as e.g. the Nile at its section in Egypt under the great dam of Aswan). Accordingly, river energy will increase and, in case of an unconsolidated bed, will rapidly incise. The extent of incision is well indicated by the fact that along the 111 km section under the Hoover Dam, the Colorado River cut 7.1 m deep within 14 years; under the Davis Dam along a 52 km section, 6.1 m deep in 30 years (Goudie 1995). Other studies claimed 2.3 m of incision in an average of 17 years for 41 dams (Babiński 1992).

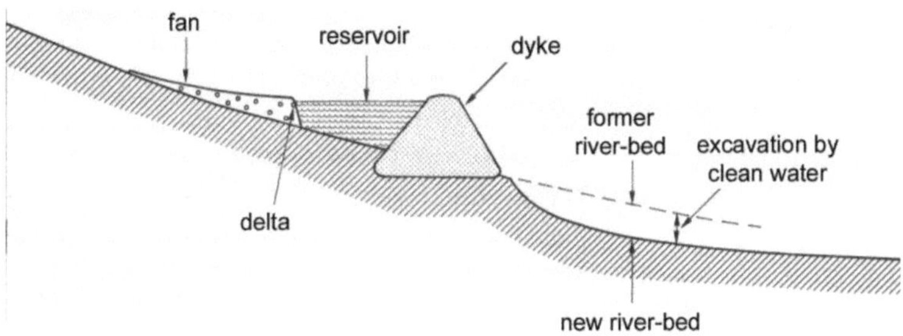

Fig. 11.3 The role of dams in the headwater and tailwater sections (Szabó in Szabó and Dávid 2006)

Since antiquity, this phenomenon took place at a number of impoundments and stepped valley floors or floodplains formed. When the damming ceased for any reasons and the reservoir became drained, occasionally even terraces formed. When damming ceased abruptly (as the dam collapsed), the streaming water and the sediment carried along not only caused disaster but rapidly transformed the morphology of the valley. The chronicle of such lake bursts in Sweden for the past 200 years was collected by Lundquist in 1944 (cited by Rathjens 1979). Most of them were the consequence of some kind of human intervention. In 1932, e.g. at the lake burst of Aovakkojaure in Lapland, water level suddenly dropped 18.5 m and downstream of the dam, a 2 km long, 100 m wide and 25–30 m deep steep-walled valley was cut. Of the coarse load flushed, the river formed a large delta at its mouth at the Luleälv River. A lake burst of similar type caused a major disaster in the Vaiont Valley in North Italy (Venetian Alps) in 1963 (for details, see Case Study 2).

It is also obvious that dammed reservoirs rapidly fill up and cannot be used any further. The filled-up lake bottom represents a new geomorphological surface. In the United States, more than 2,000 smaller reservoirs have entirely filled up until now. Accumulation can be so rapid in some cases that the benefits associated with the whole facility of huge cost can be questioned. Due to the accelerated soil erosion above the Tarbela Dam built for 9 years at a cost of 3.1 billion dollars, the lifetime of its reservoir is estimated to be not more than 20 years. The filling-up of the Lake Powell on the Colorado River is expected to happen in only 100–300 years (Chiras 1991). Reservoirs built on the Yellow River, which transports the largest amount of suspended load on Earth (920 million t per year), will fill up within 50–70 years. About 60% of the alluvium of the river is carried to the sea; however the rest, accumulated on the alluvial fan, represents such a huge volume that even one metre of loess silt could deposit on cultivated areas following a large flood.

During impoundment, at the section downstream the dam the amount of water and peak flood discharges particularly decrease. Consequently, the river bed will be narrower as well as deeper with decreasing water capacity. According to the data by Schumm (1977), the 792–1219 m in wide braided system of channels of the North Platte River at the border between Wyoming and Nebraska – following the construction of the storage reservoir – narrowed down to 60 m at some places between 1890 and 1959. For the South Platte River nearby, similar values were measured in the tailwater. The rate of river bed alterations reduces with relaxation time, as the new conditions caused by dam construction will lead to another dynamic equilibrium. This is associated by Leopold et al. (1964) with the fact that the gradient will decrease with incision and river competence also reduces (indicated by lower flood wave peaks). Therefore, the river will tend to carry only the fine (suspended) load from the bed, whereas coarse grains will remain to build a resistant armouring, which hinders further downcutting.

Flood-control dykes also have an impact on the surface. Following the construction of dykes, intensive alluviation usually begins on floodplains. A well-known example is the floodplain level of the Po River, which increased after the dam constructions by Pietro Paleocapa in the 19th century. The channelisation plan of the Tisza River by Pál Vásárhelyi, among others, was based on a concept opposite to that of Paleocapa, according to which embankments along the Tisza River should be erected closer to the river. It was believed that water flow of higher velocity on the narrowed floodplain will prevent intensive floodplain accumulation. Today measurements show that accumulation could not be avoided. In spite of the significant variation in load transport between the two rivers, its rate is much lower for the Tisza than for the Po. Floodplain alluviation is explained by the fact that a vast majority of water (above 80%) flows in the mean-stage river channel even during flood waves, whereas on the broad floodplain, there is a relatively insignificant water discharge. Therefore, the circumstances for the deposition of fine, mainly suspended load on the floodplain are especially favourable.

Straight *cut-off* sections also have a remarkable geomorphological impact. Along the cut-off section an increase of gradient takes place and will also lead to local downcutting and changes in sediment transport. These changes are apparent at sections downstream cut-offs.

Ox-bow lakes formed following cut-offs become the base levels for the immediate environs and, therefore, soon fill up. This is, however, due not only to sedimentation but also to vegetation growth during tranquil flow. Ox-bow lakes are wetlands of a special type which are of increasing ecological value in the Great Hungarian Plain, converted to a typical cultivated grassland. Groundwater flow towards the ox-bow lake can also be observed. Meander cut-offs will usually lower the groundwater table to trigger a relatively rapid transformation of soils. Areas along the River Tisza thus witnessed, at many locations, incipient or accelerated alkalisation. The specific micromorphological processes of the resultant alkaline lands, like berm formation can partially be associated with the morphological changes caused by land drainage.

Water management measures can often significantly affect sediment transport not only locally but on long river sections. Straightened channels cause a net surplus sediment transport on the reconstructed section leading to intensified accumulation on the downstream section or in the mouth. Larger amounts of sediment also reach the rivers in the wake of other, not hydrology-related, interventions. For instance, the impact of mountain deforestation increasing discharge, the reclamation of agricultural land is mentioned here. (In Hungary, the main inflow to Lake Balaton, the Zala River, doubled the amount of sediment transported to the western Bay of Keszthely from intensively cultivated areas. Sediment accumulation of various origins in many of the world's large alluvial plains resulted in measurable terrain elevations in the historic times.

The section of the increased amount of sediment reaching the sea will, in general, cause the accelerated development of *deltas*. Nowadays, however, this tendency is not common. Most of the channelisation works result in a drop of sediment load. Sediment is retained beyond large barrages in reservoirs. For the majority of the world's classic delta regions, signs of the observable degradation are observed. Silt deposition in the Nile Delta has been significantly decreased since the construction of the modern large dams (Delta Barrages – 1861, the old Aswan dam – 1898–1902, Sennar Dam – 1925, Gebel Aulia Dam – 1937, Khasmel Girba Dam – 1966, Roseires Dam – 1966, the great dam at Aswan – 1966–1972), and no delta building is recorded; on the contrary, sea transgression is characteristic. In the Rosette Branch of the delta (Fig. 11.4) archaeological finds clearly indicate that, until the Hellenistic times, a remarkable delta progression took place, settlements relocated to the north; however in the 65 years after 1898, a rather rapid degradation (1650 m) occurred (at the adjacent Damietta Branch, with a loss of 1,800 m.). A new road had to be constructed about 3 km behind the old lighthouse, and the ruins of a 19th century fortress is still found in the sea (Nir 1984; Goudie 1995). Today, even the Mississippi Delta is degraded for most of its area, mainly due to the recent flood prevention systems established along the lower river section, around New Orleans. New open-water areas appeared mainly on young marshlands (Coch 1995; Goudie 1995). Studies in Texas show that the suspended sediment transport of four rivers in Texas (the Brazos, San Bernard, Colorado and Rio Grande Rivers) dropped to 20%

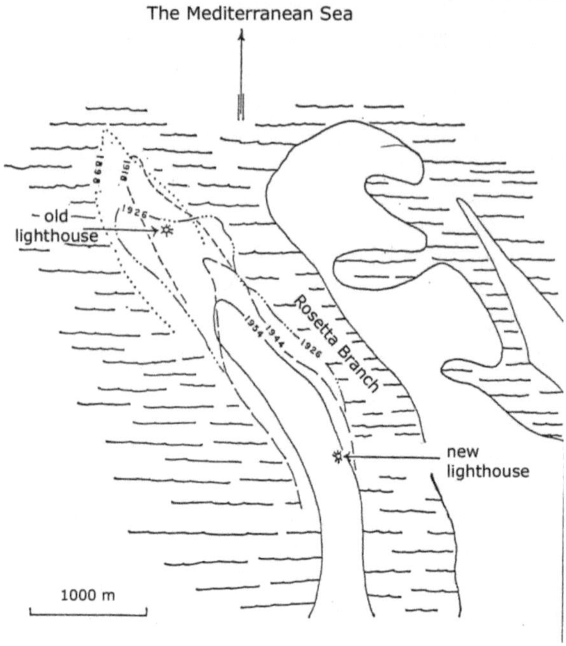

Fig. 11.4 Changes in the Rosette Branch of the Nile Delta since Hellenistic times (after Nir 1984)

between the 1930s and 1960s (after Goudie 1995). The significant decrease, in 60 years, of sediment transport of rivers flowing into the ocean at the East Coast of the USA to the south of Cape Hatteras is visible on the map (Fig. 11.5).

11.2.3 Connections between Water Regulation and Mass Movements

11.2.3.1 Impacts on Mass Movements

For mass movements, a change in at least one of their two crucial preconditions is required. Either *shear stress* has to increase or *shear strength* has to decrease. During hydraulic engineering works, both can occur but not with same frequency.

The *increase of shear stress* can primarily be a result of the increase in the load on the slope. The most obvious case is when increased slope height causes extra weight. Water management structures on slopes lead essentially to such an impact but this can be prevented by responsible planning. The *drop of shear strength* during water management is more common and can be triggered by increased slope inclination, for instance, when reservoir basins are created. This happens when canals are deepened as the slope there (the bank slope of the canal) is actually formed during construction. Such slopes may be (or appear to be) stable during construction but become steeper undercut by water flowing in the canal or by bank erosion by waves

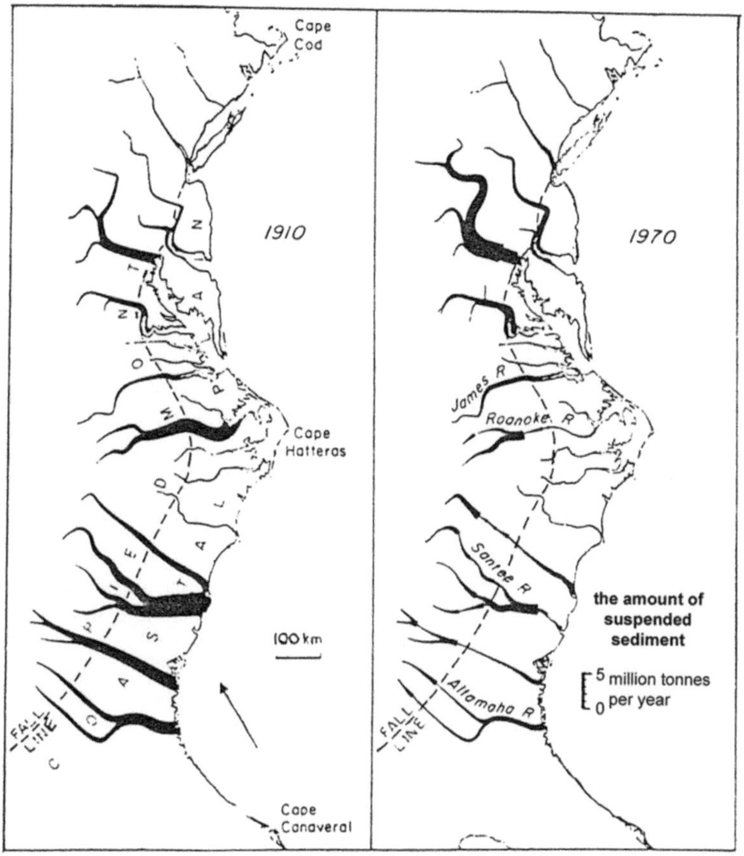

Fig. 11.5 Decrease in the sediment transport of rivers on the East Coast of the USA (after Nir 1984)

in the reservoir. Such phenomenon is the ‘wave wash’ of the dyke at the floodplain side of the embankment during the times of flood waves. This kind of mass movement coastal evolution can be regarded as common for the engineering structures mentioned. Its morphologic relevance cannot be neglected, but it rarely leads to disaster. The wave wash of the dyke, however, the can easily be the starting process for a rupture. Among the subsequent movements, falls, slides and slumpings are the most common.

Frequent and larger surface alteration (and in the meantime, greater danger) as compared to the above described is triggered when shear resistance decreases due to the changes in the slope material. Rapid changes in cohesion and angle of internal friction relevant in this respect can be caused in the simplest way by the water absorption of engineering structures (dykes, channel slopes, etc.). As such objects are usually permanently connected to water, their water adsorption can be anticipated. An obvious form of this is when the dammed-up water is pressed into the material of the bank slope due to the apparent fluid pressure. Conversely, the case can be that underground water flowing from the environment towards the channel or lake cannot emerge from the slope due to its higher water level; it is backwatered, thus decreasing its stability. Subsequently, bank mass movements can easily occur mostly when the water level of a reservoir or channel is suddenly lowered resulting in the cessation of its previous bank-pillar role. Water infiltrating the bank material can, by capillary rise, ascend even above the external water level, and can also increase soil moisture in greater elevations, thus also advancing the easier and more rapid occurrence of the saturation and movement of the bank material in the case of external water supply (e.g. rainfall).

Major morphological alterations are usually caused by mass movements. Such disasters mostly occur on the slopes of valley reservoirs. Physical processes also play a major role in triggering them; although the resultant movements are linked to anthropogenic landforms, they are considered to be semi-anthropogenic geomorphologic processes.

Most mass movements are triggered by the facilities themselves followed by the by-effect mainly also threatening them. Although, here landform alterations always take place, disaster resultant from such movements is significantly higher than their morphologic relevance. Therefore, in connection to the construction of such facilities, the necessity of professional and careful planning and implementation must be emphasised.

Case study 2: The Vaiont Reservoir disaster

The reservoir was established in the *Vaiont valley* in the late 1950s (Fig. 11.6). There are Jurassic and Cretaceous limestone strata with clayey intercalations on both valley sides, dipping towards the valley. The water of the reservoir was backwatered behind a 265.5 m high concrete *dam in the steep-sided deep valley* (Fig. 11.6). In 1960, soon after the backfilling, large-scale debris sliding was triggered on the southern slope above the reservoir. Approximately 700,000 m^3 of rock debris hailed to the lake forming waves 2 m in height. These waves running through the lake however did not cause serious damage. (As indicated by the figure on the reservoir, landslides significant in scale took place previously in the Vaiont Valley; thus, this had to be expected.) Following this, between 1961 and 1963, water level was increased and lowered 3 times with attention paid to the unsteady minor movements of the slope. In September 1963, however, so-called progressive creepings quickening initiated on the southern slope. First, the rate of movements was 3–6.5 cm per day,increasing to 20 cm per day by the end of October with more and more new ruptures being formed at the upper slope section. The occurrence of another larger slide could be taken for granted, also supported by extremely high precipitation, but experts expected a mud-rock flow occurring as in previous cases; that was forecast as relatively less dangerous. Evacuation of the surrounding areas was, however, ordered, but the landslide started after 39 minutes elapsed. It was much greater than previously expected (about 260 million m^3 – equalling to 110 Cheops pyramids – rock slid within 45 seconds), in the form of a landslide in one block. The rock mass

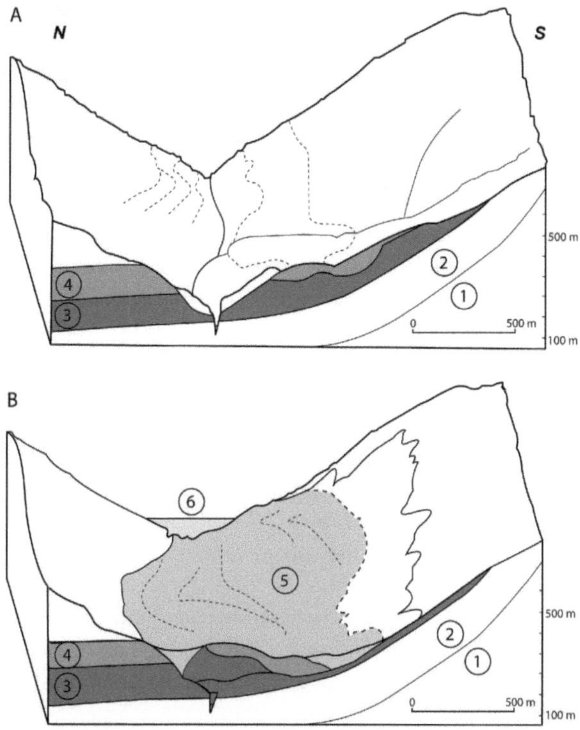

Fig. 11.6 The Vaiont Valley syncline before (A) and after the block landslide (B) (source: University of Melbourne, lecture notes). 1 – Jurassic (Lias) limestone; 2 – Dogger limestone; 3 – thinner Malm and Lower Cretaceous strata; 4 – Tertiary argillaceous sediments; 5 – slided block; 6 – impoundment level

falling into the lake containing 115 million m^3 of water generated waves exceeding 200 (!!) m in height, destroying everything on the lake shore; overtopped the dam, and streamed to the forefront from a height of approximately 500 m; ran down the Piave Valley sweeping away the settlement of Longarone; also partly Pirago, Villanova, Rivalta and Fae; taking in its wake approximately 2,000 victims. (The dam did not collapse and still stands today at the end of the former lake without water and filled by the mass of landslide. Earth movements caused by the slide were even indicated by seismographs in Brussels.) – after Coch (1995) and Petley (2001).

11.2.3.2 Water Management and Ground Subsidence

Underground water management, of which some branches, in a broader sense, can be regarded as mining is a common cause of non-slope mass movements (ground subsidence). Their relevance has become well acknowledged in the last century due to the rapidly increasing use of underground waters. Phenomena, well discussed by the media as the sinking of Venice and the increasingly frequent transgression of St. Mark's Square, drew the attention of the public. Mainly ground subsidence resultant from a volume loss due to water extraction of subsurface layers, can be associated with water management. As this is a consequence of a long-term impact, usually slow and extending gradually, its character can be, to some extent, compared to epirogenetic depressions. They are also 'related' that depressions due to water exploitation usually lack landforms unlike the mostly markable morphologic consequences of mine disasters. Consequently, depressions deriving from mining in a narrow sense (e.g. collapse of mine passages) often cause immediate disasters, unlike soil compaction unfolding often after decades. Unfortunately, water management also has negative side-effects, causing disastrous ground subsidence (collapse). Here, e.g. the more and more frequently occurring street subsidence resultant from the cracking of urban water pipelines or the collapse of buildings can be mentioned.

Within a few decades, the *extraction of groundwater* for various purposes can induce subsidence in size that can rightfully be regarded as a semi-anthropogenic natural hazard. Some data below (Table 11.1) refer to its worldwide spreading. A good example of the expansion of the subsided area associated with the increase of the value of subsidence is Tokyo where in 1960, an area of 35.2 km^2 was situated under the sea level but as a result of the near-shore groundwater extraction, in 1974 it extended to 67.6 km^2 with floods directly threatening about 1.5 million residents. Subsidence is usually the most intensive around one or some centres with its value decreasing as going further.

Table 11.1 Some examples of surface depressions due to groundwater extraction (after Goudie 1995)

Location	Value of depression (m)	Period	Rate of depression (mm per annum)
Mexico City	7.5	–	250–300
Central Valley (California)	8.53	–	–
Tokyo (Japan)	4	1892–1972	500
Osaka (Japan)	>2.8	1935–1972	76
South Central Arizona	2.9	1934–1977	96
Shanghai (China)	2.62	1921–1965	60

From this aspect, the isoline map on the subsidence of the Houston and Galveston areas in Texas between 1906 and 1978 (Fig. 11.7) is worth studying. Measurements carried out near Debrecen indicate a rather rapid descent in the context of Hungary. In this, estimated to be an annual value of 8 mm, water extraction definitely plays a decisive role, as sites indicating especially great descent fall in the areas of larger

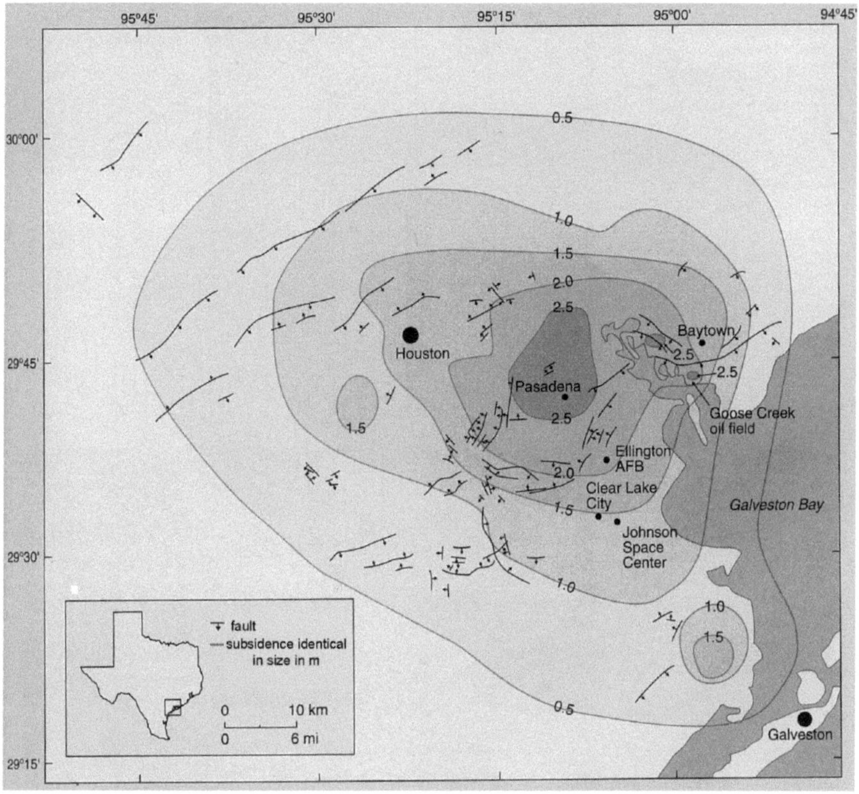

Fig. 11.7 Ground subsidence data from the districts of Houston and Galveston (after Coch 1995)

towns with a remarkable extraction of underground water (Fig. 11.8) (Szanyi 2004; Joó 1996).

Land drainage is often applied to the removal of water of areas rich in organic substances (e.g. marshes). Here, there is a remarkable decrease of the volume of draining peat and the land drainage will cause ground subsidence. In the meantime, the marsh area can be significantly compressed. It happened in the Fens of England after the draining . Between 1848 and 1957, the surface subsided by about 3.8 m and the peat area shrunk from 1750 km^2 to about 450 km^2 (data of Fillenham cited by Goudie 1995).

11.2.4 Water Construction Works and Seismicity

Accumulating experience and measurements indicate that the implementation of large-scale hydraulic engineering projects – mainly the colossal barrage dams and

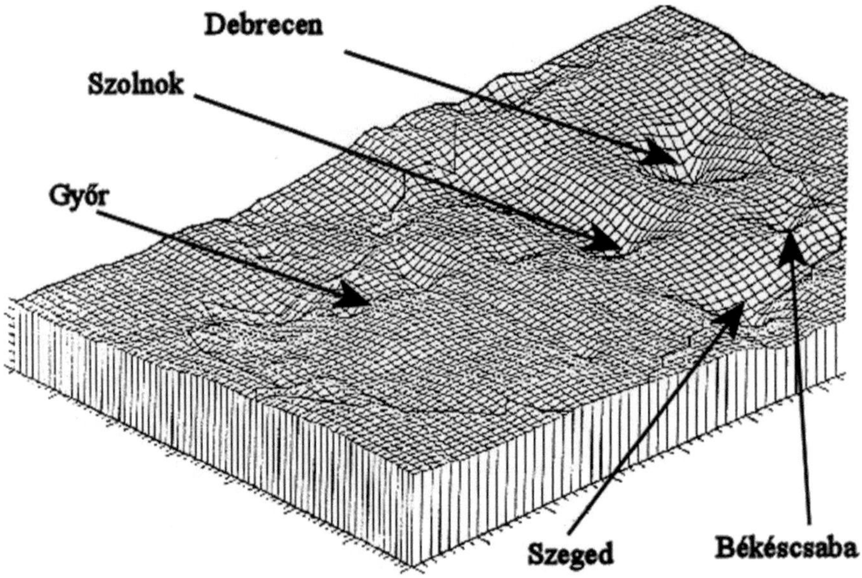

Fig. 11.8 A surface map of vertical movements in Hungary (after Joó, in: Szanyi 2004)

the water masses impounded by them – can generate underground tensions to be released in earthquakes. The most obvious assumption is that such massive constructions (or the reservoirs behind them) can generate tremors by loading pressure. It is referred by the fact that, in many cases, the frequency of earthquakes only increased following the filling-up of the reservoir. According to measurements, even the water level fluctuations of the reservoir influence the origin of earthquakes (Fig. 11.9). As pressure is increased by only about 20 bar under the deepest reservoirs, it is more likely that quakes are triggered by changes in the pore water pressure. The increased seismic activity around reservoirs fortunately has only caused relatively minor quakes until now. In the surroundings of Koyna (India), Kremaston and Marathon (Greece), Xinfengjiang (China), Kariba (Zimbabwe-Zambia) and four reservoirs at California, earthquakes exceeding 5 M Richter, related to the reservoir were observed (Clarence 1982). A close correlation between quakes and water levels has been found for a number of reservoirs.

Earthquakes can occur in the aftermath of injecting water or other fluids underground. This is only occasionally related to water management, as it is mostly carried out by mining (extraction of oil or gas). (In the late 1960s around Denver residual, nerve gas was disposed of through a rather deep well, triggering a series of earthquakes.)

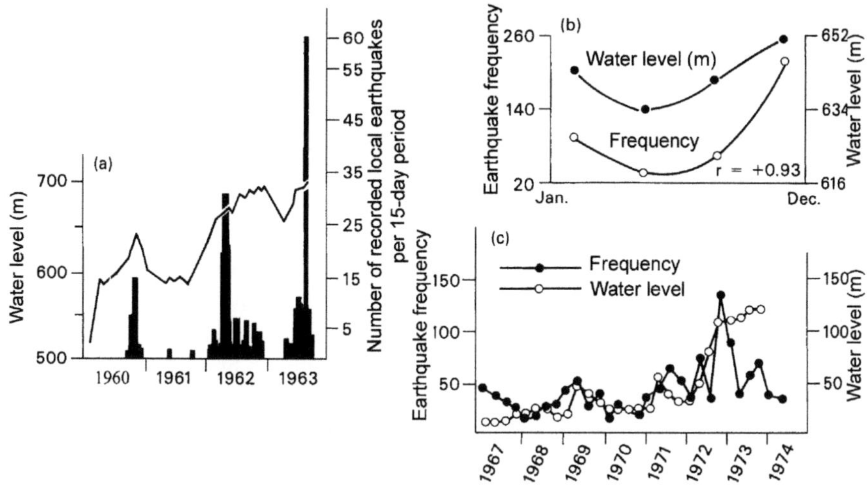

Fig. 11.9 The connection between the water level of reservoirs and the frequency of earthquakes (**a**) Vaiont Dam (Italy), (**b**) Koyna (India), (**c**) Nurek Dam (Tajikistan) (after Goudie 1995)

11.3 Coastal Interventions and Its Consequences

Frequent alterations of coasts and lake-shores have been taking place for millennia for basically two reasons: either for the purpose of *coastal defence* ('passive' intervention) or for a particular *economic activity* (transportation, expansion of cultivated land, etc – 'active' intervention). Water management forms, to some extent, part of both activities.

11.3.1 Coastal Defence

A classic example of coastal defence is when humans intend to control coastal retreat by natural processes in order to secure the population of coastal settlements. As known, *coastal retreat* derives from several causes. It is generated by the various motions of seawater (waves, tides and currents), or sea-level rise, but can also result from ground subsidence (transgression, regression). All these processes are jointly present along some coast sections in various combinations, thus making the exact identification of the causes of coastal retreat as well as the design of successful defence techniques difficult. If the causes are not properly recognised, the resultant coastal structure (as anthropogenic landforms) does not serve their purpose and may even intensify the rate of coastal degradation or undesired accumulation at various locations. In any case, human-made landforms modify natural processes and result in the development of new landforms (semi-anthropogenic processes and landforms), and this is of outstanding significance from the point of view of anthropogenic geomorphology.

Typical coastal *defence structures* are rock or concrete reinforcements of retreating cliffs, (rows of) jetties or off-shore breakwaters. At many locations in the North Sea coast of Western Europe, the chalk bluffs in England and France are protected by such structures. The constant demand for reconstruction indicates that coastal retreat cannot be halted in the long term. It is also common that degradation becomes even more intensive near coastal engineering structures. Rock revetments often cause cavity formation by stronger breakers in the foreground. Accumulation usually begins behind jetties. The beach protected by the jetty at Santa Barbara, California, built in 1930, advanced seaward more than 300 m within 7 years. Such alterations are often the consequences of wave interference induced by engineering structures. An example is the impact of breakwaters built parallel to the coastline, in order to protect harbours. Behind a breakwater of adequate length, a bar accumulates by the interference of waves arriving form both directions and after a time links the breakwater to the coast (a semi-anthropogenic tombolo bar).

Hiram, King of Phoenicia linked two islands by a dam in front of the shores of Tyrus (now Tyr in Lebanon) in the 11th century B.C. in order to create a harbour. The currents consequently altered built a tombolo bar to connect the island with the continent. In the 4th century, Alexander the Great was able to build a road 60 m in width on this bar. During the past 2,300 years the tombolo has been widened to an isthmus more than 1 km wide (after Nir 1984). A similar bar formation in front of the shores of Netanya in Israel is also mentioned by Nir (Fig. 11.10).

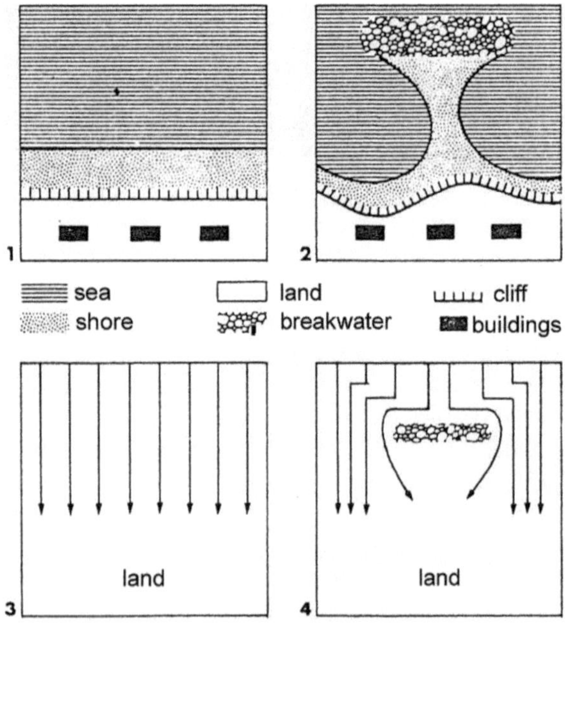

Fig. 11.10 Changes in the coastline due to the breakwater on the shores of Netanya, in Israel Pictures 1 and 3: coastline and wave directions prior to the construction of the breakwater; Pictures 2 and 4: coastline and the place of sand accumulation after breakwater construction (Nir, 1984)

The protection of the southern shore of Lake Balaton primarily aims at reducing wave erosion. Of the entire shore length of 235 km, 107.5 km is protected by embankments (data of Kleb).

11.3.2 Land Reclamation for Economic Purposes

The establishment of *ports* has been a typical example of active coast formation worldwide for a long time. The primary aim was secure anchoring, but it was accompanied by the unwanted consequence of coastal retreat at many locations. Deliberate land reclamation can be achieved not only by creating artificial islands or landfills but also by influencing coastal water motion over larger areas to result in coastal advance. Not surprisingly, the techniques developed derive from countries with low or subsiding coasts endangered by flooding or where the surf in shallow off-shore waters transported large amounts of material towards the shore allowing land reclamation by its retention and fixation.

11.3.2.1 Land Reclamation on the North Sea Coast

Classic examples of land reclamation are found on the Dutch–North German–Danish shores and also longer sections of the British Isles. In the background of the Wadden Sea actual tidal zone, several kilometres in width, multiple *brushwood palisades* with gates were planted. The deposits reaching the shore are retained by the wooden structures continuously raising the surface. This primitive water management technique was one of the earliest successful ways of polder formation. The method of enclosing marshlands at or below sea level by dams has been widely applied for ages. The surface- and groundwater springing due to the hydrostatic pressure were pumped over the dykes and conducted away. Windmills proved to be useful in this respect in coastal areas with frequent strong winds. Making such lands arable and maintaining cultivation on them was only possible by well-organised water regulation systems (hierarchic canal systems, sluices). For dyke constructions, people took advantage of coastal dune ranges of natural origin. The gaps were blocked and the dunes themselves were not only raised by accumulating material upon them but by promoting vegetation growth on their surfaces. This latter activity can also be considered a semi-anthropogenic geomorphic process, as humans took advantage of natural processes for their purposes. A well-known fact is that polder formation started according to the ambitious plans of the Dutch engineer Cornelis Lely and peaked in the Modern age in the about 50 years after 1920, leading to the development of most of the polders in the country. Following the efforts of past centuries, it significantly contributed to the formation of The Netherlands' face. A modern continuation of the classic coastal protection is represented by the implementation of the Delta Plan (1953–1978) providing a new character to especially the Seeland Region. Water management in The Netherlands is obviously a key economic branch and affects all fields of life but primarily and basically the surface.

This is well symbolised by the fact that the Water Ministry of the Netherlands is called Waterstaat, indicating the close contacts between water management and the State. The achievements of land reclamation in the Netherlands since the 10th century are known in detail since 1540 and affect approximately 610,000 ha in total (Fig. 11.11).

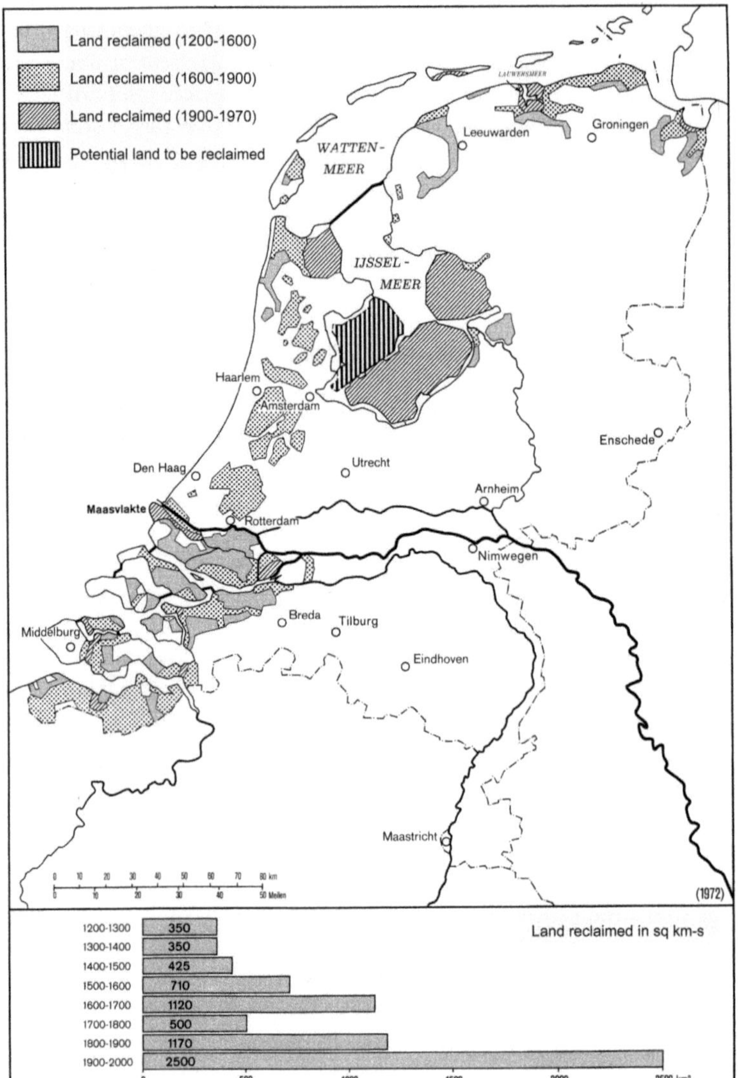

Fig. 11.11 Polder formation in the Netherlands from the Middle Ages to the 1970s (after Informations und Dokumentationszentrum 1975)

In several coastal zones of the British Isles, especially at the estuaries of Humber, Thames and the Wash, as well as in Kent (Romney Marsh), the extent of newly reclaimed lands since the rule of James I (1640) is estimated to be 525,000 ha. In France, in Normandy (at the Isle of Mount St. Michel) in the Loire estuary region, lands requiring study by anthropogenic geomorphology were created.

References

Alföldi L (2000) Árvíz, belvíz talajvíz (Flood, excess water, groundwater). Magy Tud 107: 673–687

Babińiski Z (1992) Hydromorphological consequences of regulating the lower Vistula, Poland. Regul Riv: Res Manag 7(4): 337–348

Chiras DD (1991) Environmental Science – Action for a Sustainable Future. The Benjamin/Cummings Publishing Company, Redwood City

Clarence RA (1982) Reservoir-Induced Earthquakes And Engineering Policy. Reprinted from Proceedings of Research Conference on Intra-Continental Earthquakes, Ohrid, Yugoslavia, 17–21 September 1979

Coch NK (1995) Geohazards – Natural and Human. Prentice-Hall, New Jersey

Der Fischer Weltalmanach 2002 (2001). Frankfurt am Main

Goudie A (1995) The Human Impact on the Natural Environment. Blackwell, Oxford

Informations- und Dokumentetionszentrum für die Geographie der Niederlande (1975) Zuidersee/Ijsselmeer. Den Haag

Joó I (1996) A földfelszín magassági irányú mozgásai Magyarországon (Vertical movements of the ground surface in Hungary). Geod és Kart 1996 (4)

Kollmorgen WM (1953) Settlement Control Beats Flood Control. Econ Geogr 29 (3): 208–215

Lászlóffy W (1982) A Tisza (The Tisza River). Budapest

Legget RF (1962) Geology and Engineering. 2nd edn. New York

Leopold LB, Miller JP, Wolman MG (1964) Fluvial Processes in Geomorphology. San Francisco

Mock J (1992) Auswirkungen des Hochwasserschutzes. In: Böhm HR, Deneke M(eds.), Wasser – Eine Einführungin die Umweltwissenschaften. Darmstadt

Nir D (1984) Man, A Geomorphological Agent. Keter Publishing House, Jerusalem

Petley D (2001) The Vaiont Landslide. http.www.land-man.net/vajont/vajont.html

Rathjens C (1979) Die Formung der Erdoberfläche unter dem Einfluβ des Menschen. Teubner Studienbücher, Stuttgart

Schumm SA (1977) The fluvial system. Wiley, New York

Szabó J (1993) A társadalom hatása a földfelszínre (The impact of society on the Earth's surface). In: Borsy Z (ed.), Általános természetföldrajz. Nemzeti Tankönyvkiadó, Budapest

Szabó J, Dávid L (eds.) (2006) Antropogén geomorfológia (Anthropogenic Geopmorphology) University notes. Kossuth Egyetemi Kiadó, Debrecen (the Hungarian version of this book)

Szanyi J (2004) Felszín alatti víztermelés környezeti hatásai a Dél-Nyírség példáján (Environmental impacts of underground water exploitation in the Souther Nyírség Region). Manuscript PhD Thesis, University of Szeged

Tóth Cs, Félegyházi E, Szabó J (2001) The study of the middle Tisza region dead riverbeds from the aspect of landscape rehabilitation. In: Buzek L, Rzetala M (eds.), Man and landscape. Ostrava-Sosnoviec

Chapter 12
Urban Development and Anthropogenic Geomorphology

Péter Csima

Abstract The spreading of settlements and urbanization processes, along with the construction of the necessary infrastructure are of decisive local influence on geomorphic evolution. The chapter investigates the following general problems: What are the most important impacts of settlement development on geomorphic evolution? What environmental principles have to be observed during urban planning and the implementation of development plans? What are the most typical human-induced processes and surface features in settlements? What consequences anthropogenic intervention has on the land use, environmental conditions and scenery of settlements? Among the aesthetic consequences and positive impacts, artificial mounds in green areas and artificial water surface are mentioned, while among the negative influences the visual appearance of flood-control dykes and the large-scale road-cuts and supporting walls are listed.

Keywords Urban development · Landscaping · Urban anthropogenic landforms · Terrain transformation

12.1 Introduction

Interactions between relief and urban development have been well known for centuries. Landscape, as one of the 'allocating factors' (Mendöl 1963), was a major consideration for the site selection of settlements. Detailed research into its multifaceted impacts has always belonged to the traditional topics of urban geography. The impact in the opposite direction, i.e. of expanding settlements on the topography of their environs, however, received relatively less attention. In the course of urban development, modified relief takes a modified path of evolution. It means that

P. Csima (✉)
Department of Landscape Protection and Reclamation, Corvinus University of Budapest, Villányi út 35–43, 1118 Budapest, Hungary
e-mail: peter.csima@uni-corvinus.hu

J. Szabó et al. (eds.), *Anthropogenic Geomorphology*,
DOI 10.1007/978-90-481-3058-0_12,

in addition to natural topography, anthropogenic landforms should also be taken into account, and this subject is regarded to be essential in anthropogenic geomorphology from both a conceptual and a practical point of view. The issue is of particular significance for developing countries (Gupta and Ahmad 1999).

This chapter discusses

- the impacts of urban development on topography,
- the principles of settlements' terrain correction planning and implementation,
- characteristic anthropogenic landforms in settlements,
- the impacts of geomorphic processes in settlements on land use, ecological and urban conditions.

12.2 Geomorphic Impacts of Urbanization

The most characteristic geomorphological consequences of urban development are the following:

Modified drainage and surface sealing change or stop previously operating natural geomorphic processes.

New surface is developed next to buildings and other structures (terraces, slopes, drainage facilities, artificial ponds, etc.).

Debris and waste accumulation results, over a longer period of time, in landfills (as in the area of the Castle Hill of Buda).

The majority of minor features are destroyed by construction works, which can also be a direct threat to geomorphic values (e.g. Budai Hills).

Particularly in town centres and in industrial-commercial areas, most of the ground surface becomes built-up or sealed and, consequently, runoff conditions are modified.

Unfavourable geomorphologic processes, such as erosion on slopes or mass movements emerge (Szabó 1993).

Artificial hollows developed during urban engineering can cave in (Gálos and Kertész 1997). Examples from Hungary are the cellars in the centre of Eger and Pécs.

Landfill in natural and artificial depressions by construction rubble and materials excavated during the construction of fundaments.

During settlement development, the same surfaces can undergo multiple alterations. The construction of trenches and ramparts for strategic defence purposes is mostly also linked to towns. Some of them were part of fortifications, whereas others were built along the city borders. Most of them, however, had been abandoned by urban development by the 21st century, and become sites for anthropogenic surface formation for a second time (e.g. Szolnok, Komárom, Sárvár).

The construction of dams for flood prevention created a new situation in the 19th century when built-up areas extended over former floodplains (e.g. in the case of Szeged on the Tisza River).

A special case of geomorphic impact is related to the extraction of building materials, and, subsequently, to placing and extending buildings into the slopes at the rear of yards in loess hills. In certain villages, these cuts have become decisive elements of settlement structure or of the landscape (e.g. at Gomba and Bénye in Pest County and at Gyönk in Tolna County, Hungary – Fig. 12.1). Surface alterations of a similar scale also took place along the margins of the North Hungarian Mountains, where the patterns and views of villages are also defined by walls, cellars and cave dwelling cut into volcanic tuffs (e.g. in Noszvaj and Cserépfalu in the southern foothills of the Bükk Mountains; Egerszalók, Sirok and others between the Bükk and the Mátra Mountains).

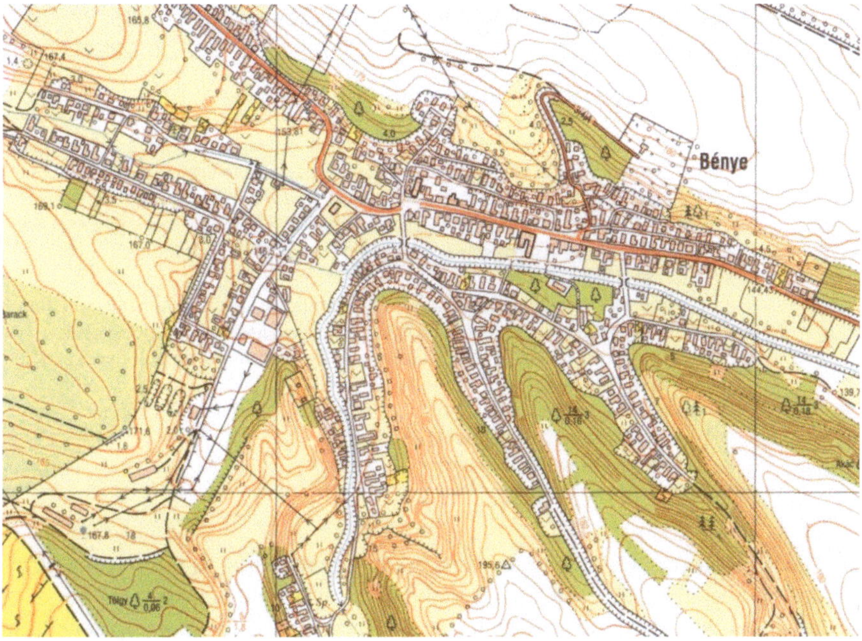

Fig. 12.1 Small settlement cut into high loess bluffs, Bénye, Pest County, Hungary

12.3 Settlement Infrastructure and Topographic Transformation

The surface character has an influence on the arrangement and inner structure of infrastructural networks and structures such as the road network, railway lines and public utilities in settlements. The role of infrastructure elements in man-induced geomorphic transformation is also significant (as in the case of railway embankments, airports, ground drainage networks). As the character of surface transformation is similar to the impacts of the construction of infrastructure outside the settlements, they are not discussed in detail here.

12.3.1 Development and Terrain Transformation as a Function of Slope Angle

In addition to other geomorphologic factors, the angle of slope also affects the identification of potential sites for urban development. Based on literature data, planning guidelines and principles, the table below summarizes the relationship between the mean angle of slope, housing development and landscaping (Table 12.1).

Table 12.1 Mean angle of slope, housing development and landscaping

Angle of slope	Development potential and the required landscaping
Up to 5%	Area with easy and economic development potential. In general, terracing is not necessary; landscaping exclusively restricts to drainage. Relief does not mean a limit either to build-up density or building sizes.
5–12%	Increased development costs. Landscaping is inevitable; development is only possible with terracing and slope levelling. Somewhat limited development.
12–25%	Area with a development potential at significant cost and labour expenditure, only with terraces and supporting walls provided. Major topographic transformation; the relief fundamentally controls the type of development to be applied.
25–35%	Surface with limited potential for urban development. Low building density with small-sized buildings allowed.
Above 35%	Surface unsuitable for urban development.

The percentages indicated in the table are only guidelines, as the housing development potential of a given area greatly depends – among other factors – on bedrock stability, the joint pattern on the surface, the building size planned and the parameters of the access road network. It is also important to note that such limitations mainly apply to contiguous urban-type architecture. This does not necessarily exclude the siting of individual buildings in slope categories unfavourable for urban development. Today technological progress is not limited by slopes; development, however, often results in large-scale aesthetic destruction and degradation of ecological conditions.

Specific costs of construction works can significantly increase due to the necessity for landscaping. This, however, would not scare off those intending to build on steep slopes, as the value of estates is highly increased by favourable relief. A fine panoramic view is a significant factor of value added.

The above remarks apply only to natural relief conditions. Even more attention should be paid to them when further development takes place on artificially shaped slopes. There may be a high variation in stability when compared to natural slopes with similar parameters.

12.4 The Principles of Urban Landscaping, Its Planning and Implementation

12.4.1 The Need for Landscaping

As part of architectural/landscape architectural activities, physical landscaping includes the intentional and expedient modification of surfaces (Ormos 1953). Landscaping is based on technical planning. The transformation of the terrain for construction is also meant to impede surface erosion and mass movements. The steeper a slope is, the more jointed the surface is, the larger the scale of landscaping required.

During macro-scale landscaping, level or gently sloping surfaces have to be established either by planating rough surfaces or by terracing or slope levelling. From a geomorphic aspect, levelling means 'obliteration of landforms' whereas terracing means 'creation of landforms' (Chapter 10).

12.4.2 Macro- and Micro-scale Landscaping

Landscaping associated with urban development is practically classified by the degree of its geomorphic impact into categories of micro-scale and macro-scale landscaping. *Macro-scale landscaping* affects large surfaces like plots for housing blocks, industrial areas and sports facilities before development.

Micro-scale landscaping, on the other hand, refers to smaller surfaces like (parts of) plots within individual properties.

Since they differ in their different functions, it is important to define the goals of landscaping for architectural and landscape architectural (landscape management and gardening) purposes. For *architecture*, macro-scale landscaping facilitates the placement of buildings and prevents mass movements. In *landscape management,* macro-scale landscaping promotes the restoration of disturbed water budget (Chin & Gregory 2009); allows landscape rehabilitation (the elimination of scars or destructed surfaces that emerged from either previous activities or from construction works); fits architectural ensembles into the landscape from an ecological and landscape-aesthetic aspect – facilitating construction while preserving the original character of the surface.

For *architecture,* the reason for micro-scale landscaping lies in preserving the condition of buildings (for instance, through draining away excess water) and in fitting buildings and structures into the urban landscape (street view).

For *garden architecture*, macro-scale landscaping aims at improved drainage (usually forming an artificial recipient, a canal or a ditch); at preventing surface erosion; at facilitating the establishment of green areas, parks and gardens as well as habitats favourable for plants and at providing an optimal ecological and aesthetic environment for buildings and structures, through promoting favourable interactions between the original and artificial surfaces.

12.4.3 Planning and Implementation Objectives

In the landscape design phase of urban development plans, surfaces of natural value and artificial surfaces with cultural–historical values in the planning area have to be identified (Csima et al. 2004). The measures required for their conservation or possible restoration have to be specified.

In the environmental and green area management phase of urban development plans, the principles of landscaping are defined and numeric values can be prescribed as far as the permitted height, slope gradient and other parameters of earthworks in the construction area are concerned.

Landscaping techniques have to be designed as part of the environmental arrangement (garden architectural) work phase, in accordance with the directives of building measures (in Hungary, primarily *National urban development requirements*). Planning can be carried out by providing the altitude data and gradients proposed or by contour representation. The activities related to earth excavation, removal and accumulation at other locations are called 'earthworks' by the technical nomenclature and can be 'linear, areal or hollow in character' (Markó 1975).

12.4.4 Typical Anthropogenic Landforms in Urban Areas

Urban anthropogenic landforms resulting from landscaping can be studied on two levels:

- the level of urban architecture and
- the level of construction of individual buildings and structures.

The main groups of anthropogenic landforms influencing both settlement pattern and the urban landscape are

- earthworks like ramparts, mottes for strategic defence purposes, (for instance, the fortifications in the town of Komárom, being decisive for settlement pattern) protected as local historical values;
- macro-terraces, backfillings or plateaux for housing development (residential blocks, terraced houses, streets; as in Salgótarján or Miskolc in Hungary);
- water management facilities including flood-control dykes, with particular prominence in lowland Hungary (in Komárom, Szolnok, Szeged), regulated river channels (the Élővíz Canal at Gyula), banks of rivers or streams (in Budapest,

but Szentendre and Veszprém with a more varied topography are also good examples), lake shores (in Siófok, Gárdony), surfaces upfilled for the purpose of flood prevention (in Szeged);
- surfaces of transportation facilities, among them, linear embankments of national main roads and railways, airports, noise prevention mounts.

The anthropogenic surface features which are part of architectural facilities can be funtionally classified as follows:

- terraces, slopes, supporting walls and surface drainage systems developed for building safety;
- green surfaces above underground facilities (such as underground garages);
- earthworks of local public utility works (e.g. water reservoirs) and of local transport facilities;
- landforms related to garden architecture: sledging mounds (in the housing estates of Budapest) (Plate 12.1), outlook platforms (on the Gellért Hill, Budapest), large mounts for bordering or dividing space (in Kecskemét and Szekszárd), surfaces of sport facilities, garden supporting walls and slopes, garden ponds.

Plate 12.1 Sledging mound in a public park in Budapest (Csima 2008)

12.4.5 Secondary Impacts of Surface Transformation at Settlements

Urban landscaping, i.e. the establishment of anthropogenic landforms can also influence land use and the ecological and aesthetic properties of the settlement in question. Within each group, positive and negative impacts are distinguished.

The impact on land use is positive if landscaping makes a given area suitable for development and negative if it results in subsidence, collapse, eventually making the area unsuitable for construction. The ecological consequences are positive if new habitats emerge (in gardens and urban parks) and negative if former habitats disappear in the wake of alterations in water budget and solar radiation conditions (e.g. on partly built-up hillsides). A favourable change in aesthetic conditions results from landscaping if artificial mounts established in green areas successfully divide space and if the water surface of artificial channels becomes more spectacular. A feeling of oppression is created by masking river embankments and by establishing major cuts (slopes, supporting structures) (Plate 12.2).

Plate 12.2 The building of high-rise residential homes involves a major transformation of the terrain (Csima 2008)

References

Chin A, Gregory K (2009) From Research to Application: Management Implications from Studies of Urban River Adjustment. Geogr Compass 3(1):297–328

Csima P, Gergely A, Kiss G, Módosné BI (2004) Természetvédelem – védett területek tervezése (Nature conservation – Planning protected areas). University Notes, Budapest University of Economics and Administration, Budapest
Gálos M, Kertész P (1997) Mérnökgeológia (Engineering geology). University Notes, Műegyetemi Kiadó, Budapest
Gupta A, Ahmad R (1999) Geomorphology and urban topics: building an interface between research and usage. Geomorphology 31(1–4):133–149
Markó I (1975) Földművek – védelem (Earthworks – prevention). Műszaki Könyvkiadó, Budapest
Mendöl T (1963) Általános településföldrajz (The geography of settlements). Akadémiai Kiadó, Budapest
Ormos I (1953) Kertészeti földméréstan és kerttechnika (Horticultural geodesy and techniques). Mezőgazdasági Kiadó, Budapest
Szabó J (1993) A társadalom hatása a földfelszínre (The impact of society on the surface). In: Borsy Z (ed.), Általános természeti földrajz. Egyetemi tankönyv, Budapest

Chapter 13
Transportation and Industry

Lóránt Dávid, Zoltán Ilyés, and Zoltán Baros

Abstract Alterations in the topography due to the construction of transport infrastructure and industrial development are rather complex processes. The impact of transport constructions upsetting (topographic) equilibrium is manifested in relatively narrow strips and mostly, through producing abnormally steep slopes, in reducing relief stability. The earthworks for transport routes are themselves geomorphic factors whereas in the case of industrial developments, planation is usually influential. Topographic changes related to the construction of transport infrastructure and industrial development are discussed historically in this chapter. First, direct impacts related to the construction of Roman and Medieval roads, hollow roads in loess, public roads, motorways, railways, canals, tunnels and airports are discussed, followed by impacts of early mining and metallurgy, cellars, sludge reservoirs, slag cones and fly-ash reservoirs, cooling ponds, industrial parks, shopping centres and waste disposal sites. Among indirect impacts, an introduction is given to the consequences of surface sealing, changes in runoff, the 'waterfall effect' as well as to environmental impacts under permafrost conditions.

Keywords Transportation infrastructure construction · Direct impacts · Indirect impacts · Modern industrial development

13.1 Transport Infrastructure Construction and Industrial Development

Interventions with topographic impact involved by the construction of transport infrastructure and industrial development are rather complex processes, consequently they might be linked to or overlap with other chapters of this book (with water management, urban processes, mining). The impact of transport constructions upsetting (topographic) equilibrium is manifested in relatively narrow strips and the abnormally steep slopes created mostly reduce relief stability. The earthworks for

L. Dávid (✉)
Department of Tourism and Regional Development, Károly Róbert College, 3200 Gyöngyös, Hungary
e-mail: davidlo@karolyrobert.hu

J. Szabó et al. (eds.), *Anthropogenic Geomorphology*,
DOI 10.1007/978-90-481-3058-0_13,

transport routes are themselves landscape-forming factors and, in addition, indirectly influence geomorphic and microclimate-forming processes (Erdősi 1987). In the case of industrial developments, planation is usually mentioned. Changes in the anthropogenic relief related to the construction of transport infrastructure and industrial development are discussed below in a historical order.

13.2 Transport and Industrial Infrastructure Until the Modern Age

13.2.1 Transport Routes

Among the early transport routes, from the point of view of geomorphology, *Roman roads* are the most enduring structures. They entailed, before large-scale motorway constructions of the 20th century, the removal of the most significant amount of material. The technique applied for old gravel roads widely used during the fourth to fifth centuries B.C. was rather simple: the foundation was first tamped and then spread with gravel. Paved roads with concrete surface were first built around 400 B.C. The longest and best known is the Via Appia built from 312 B.C.; it has become the finest example of Roman road construction for centuries. In Western Hungary, sections of the Amber Road have been preserved and are still well observable.

The construction technology of paved roads is complex: first, forests were cleared in strips 60 m wide along the roads followed by digging drainage ditches in a distance of 12–15 m. Earth excavated from such fossae were piled in dykes (aggers) ca. 1 m in height securing the road.

Roman roads had a layered construction. (The terms Strasse and street also originated from the Latin via strata, or layered road.) Onto the tamped clay, 25–60 cm high quarry-stones were first placed (statumen), followed by a cemented layer of smaller, big clump-sized rocks (rudus, ruderatio). Rudus was followed by the nucleus that might have contained walnut-sized crushed rocks, gravel, coarse sand and carbonate debris. The final layer was the road surface or cover (summa crusta, pavimentum, summum dorsum) consisting of ca. 60 × 60 cm, 25 cm thick, mostly volcanic flagstones (Fig. 13.1). Sloping road surfaces ensured runoff towards the edges. Of tamped clay or sand, sidewalks were also constructed. Along the roads, milestones, on which the distance of the nearest town was indicated by Roman miles (ca. 1.48 km) were placed. When completed, the roads rose as high as 2 m above the surface. In the summa crusta, wheel-tracks are often seen. Freeze–thaw alteration and scuffing of the gravel necessitated maintenance: Roman roads were completely restored after ca. 100 years of use (Klischat 1996).

In the 1st century A.D., the military significance of roads declined whereas comfort aspects became more important, thus that time witnessed a recursion to the construction of gravel roads on which coaches could run smoothly. The other reason for this change of technology was the fact that most Roman provinces were lacking

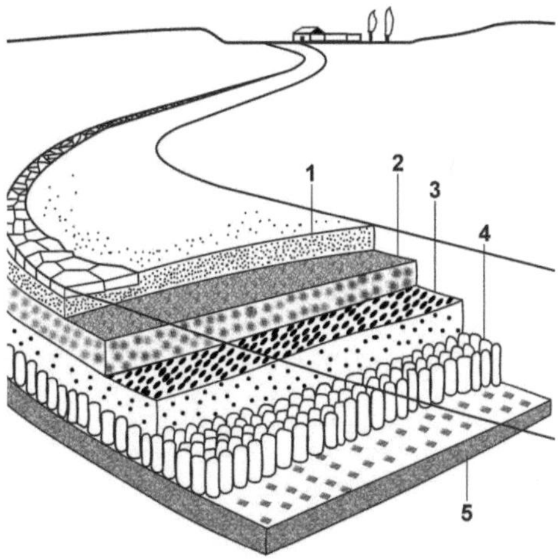

Fig. 13.1 The structure of Roman roads (**1**: summa crusta – gravel sand or stone pavement, **2**: nucleus – walnut-sized stones, **3**: ruderatio – fist-sized boulders, **4**: statumen – quarry-stones and binder, **5**: – tamped clay, the roadway is ca. 1 m thick) (re-edited after Klischat 1996)

volcanic rocks required for paving. Road-cuts were applied, minor tributary valleys were bridged by embankments, viaducts, depressions were filled and tunnels, even several hundred kilometres in length, were dug. Roman roads of straight alignment formed a network in the Empire just after Christ.

Ample morphological evidence of transportation routes has survived from the Middle Ages. The majority of medieval roads, unlike Roman roads, were not surfaced; at some places, however, traces of gravel spreading and debris fill can still be found. Medieval roads, usually 4–9 m in width, often followed watersheds on hill and mountain ridges, averted watercourses and waterlogged areas, marshlands and gallery forests of valley floors. In mountainous regions, road-cuts into the bedrock can be identified. Traces of the formation of sunken or hollow roads are also common along medieval roads. *Sunken roads* were classified in Germany by Dietrich Denecke, and he also developed a methodology to study the morphology, formation and dating of road tracks (Denecke 1969). The most important physical factors of sunken road formation include the angle of slope, soil, bedrock and vegetation. Sunken road formation is a type of gully erosion determined by the angle of slope. The mechanics of rocks and soils greatly influence the development and preservation of sunken roads. The incision of wheel-tracks can be extremely rapid on loess, loamy soils and banked rocks. In the sandy and clayey floor, sunken roads are less capable of preserving their shape, and take on a bowl shape. The lack of vegetation, on the one hand, contributes to an accelerated deepening of roads as well as to the further rill or gully erosion following abandonment. The damage caused by the recent use of roads, wheels and tramping, depending on intensity, hinders the formation of grass or tree cover and promotes deepening. Grass and forest vegetation, on the other hand, help preserve sunken road profiles and the identification of fossil sunken roads (Denecke 1969).

Two types of active sunken roads are distinguished by Denecke: wheel-tracks and trapeze-shaped sunken roads. According to him, the fossil type includes sunken roads of rounded profile without sharp edges, V-shaped sunken road traces cut deep into the less resistant material bordered by steep slopes and accumulated, planated, wide-floored sunken roads. Relict landforms as terraces and dells evidence abandoned sunken roads. Some of the sunken roads of the Modern age have been paved (Denecke 1969) (Fig. 13.2).

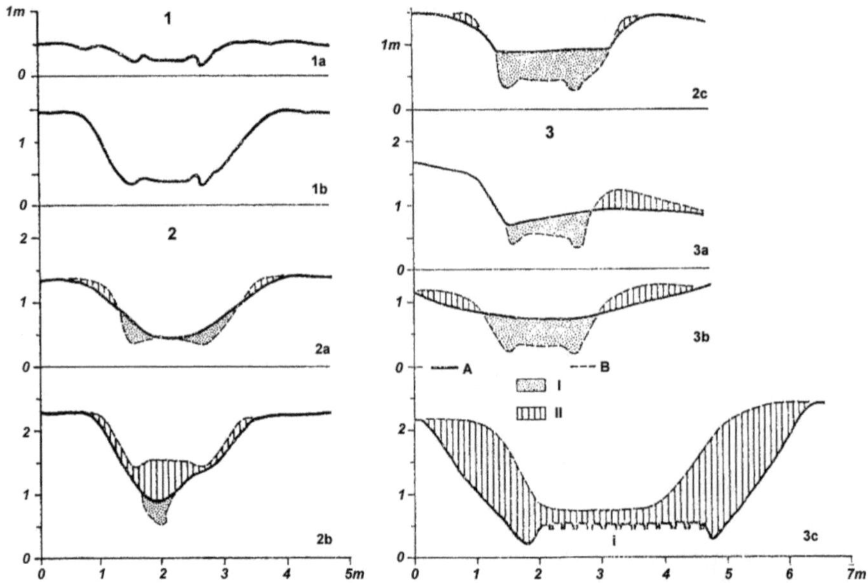

Fig. 13.2 Ideal profiles of sunken road types (**1**: Recent landform types: **1a**: wheel-track, **1b**: trapeze-shaped sunken road, **2**: Fossil landform types: **2a**: trough-shaped sunken road, **2b**: sunken ravine, **2c**: wide-floored sunken road, **3**: Relict landform types: **3a**: sunken road terrace, **3b**: sunken road dell, **3c**: paved sunken road, A: present-day profile, B: earlier profile, I: removed material, II: eroded loose material, i: stone or gravel cover) (Denecke 1969)

Hollow roads in loess are typical erosional landforms with a U-shaped cross-section (Gönczy and Szalay 2004). They are dirt-roads where the primary loess structure is crushed by vehicles to dust. Their development is closely related to carbonate content, the capillary structure of loess or sandy loess as well as to gully erosion. Wheel-tracks rainwater runoff, especially during heavy rainstorms, entrains a large amount of material and gradually deepens the roads. After decades, the former roads are gradually transformed into hollow roads with (sub)vertical walls. Their depth may range from 2 to 10–15 m, in China even to 40 m (Plate 13.1, Borsy 1993).

Hollow roads in loess, as a consequence of piping and gully erosion, are transformed into steep-walled, V-shaped ravines (Gönczy and Szalay 2004). Thus, further dirt-roads have to be made on cultivated land (Borsy 1993). It is a typical situation in Hungary, especially in the counties of Fejér and Tolna and along the eastern rim of the Mezőföld Plain (Szilágyi 2001), in the Solymár, Pilisborosjenő and Üröm basins (Buda Hills) and in the southern foregrounds of Hosszú Hill (Rübl 1959). The widest range of loess denudation landforms (among them, hollow roads) in the country are found in the Szekszárd Hills (Ádám 1964 and Endrédi 2001) but are also typical at other parts of the Carpathian Basin (e.g. Szalánkemén in Serbia; Plate 13.2).

Plate 13.1 Hollow road in loess in China (from the book 'The Chinese Empire' by L. Lóczy, cited by Borsy 1993)

Plate 13.2 Hollow road in loess near the village of Szalánkemén along the River Danube (Dávid 2005)

13.2.2 The Early Impacts of Industry

Roads of greater length were first constructed on embankments as well as more remarkable road-cuts created by explosion were first applied in the Modern Age. From the late 18th century, main, mostly military, roads were again paved. Straight-aligned paved roads, chaussées and roads in major principal centres of the Napoleonic times are examples (Rathjens 1979).

Early industrial sites are inseparable from mining (e.g. in the Harz and Ore Mountains in Germany and near the mining towns of Upper Hungary). In regions of ore mining and metallurgy, there are traces of manual techniques of early mining, mine adits, waste heaps of various shape, ore-mills and washeries, smelters and foundries, slag piles, traces of coal burning, dams of reservoirs and remnants of settlements. The industrial heritage from the era prior to the Industrial Revolution can often be studied in archaeological explorations and are less relevant for geomorphology.

In addition to mining and metallurgy, many other industries shape the surface. The exchange of products and raw materials called for a dense road network. The settlement of smelters was usually determined by the distribution of plants for charcoal production used for metallurgy on a large scale. In areas of higher relief, charcoal and ore transportation roads locally turned into sunken roads. The large-scale production of charcoal used for metallurgy caused typical alterations in the relief: terraces are still identifiable at sites of former wood piles after centuries (Fig. 13.3). Molehills containing charcoal also refer to the places of wood piles. With increasing demand for charcoal, early coal burning in pits was replaced by larger scale burning in wood piles resulting in larger coal masses from the 14th century.

Former metallurgy is evidenced by the remains of stamp mills and ore-smelting furnaces, charcoal storages sites and smaller or larger slag piles containing the by-products of metallurgy (Willms 1998). Slag piles are often hardly visible now, vegetated with heavy metal-tolerant plants and adjusted into the landscape and are visited by mineral collectors (Fig. 13.4).

For stamp and grinding mills, foundries and smithies, the proximity of water was essential: streams were impounded or water drained from mines was captured by artificial embankments in order to increase their kinetic energy. In the 18th century, in Selmecbánya (now Banska Štiavnica in Slovakia), a system with a water volume capacity of 7 million m^3, according to the plans of the innovative engineer and cartographer, Sámuel Mikoviny, comprised not only of water reservoirs, but approximately 60 km of impoundments and 35 km of drainage ditches. The dams were built of tamped clay and their sides were later covered by quarry-stones to prevent erosion (Ilyés 2004). Following the Industrial Revolution, with the advent of modern manufacturing, unprofitable, sporadic foundries were gradually closed down. Their sites are mainly indicated by relief features at the sites of dams and channels, and remains of buildings. The Central European town Mecenzéf (now Medzev in Slovakia), before World War II, had the highest number of foundries (140) operating for the longest period of time.

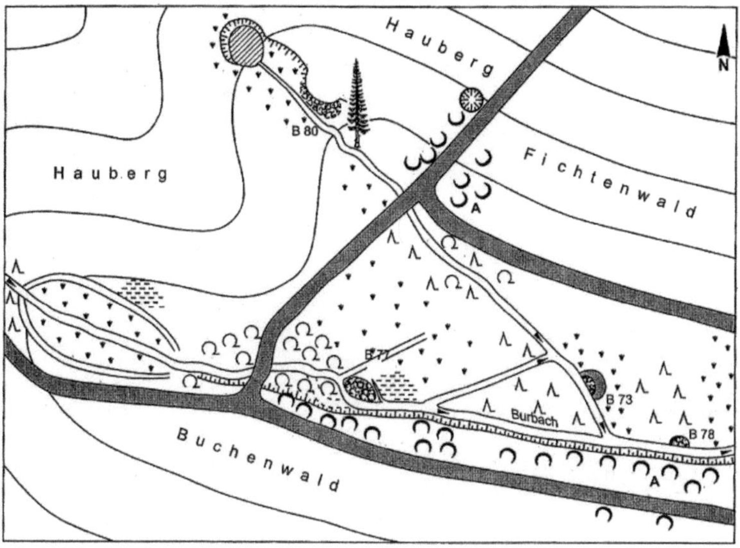

Fig. 13.3 Traces of coal burning in a mining-industrial area of Central Germany (A: site of the wood pile) (Willms 1998)

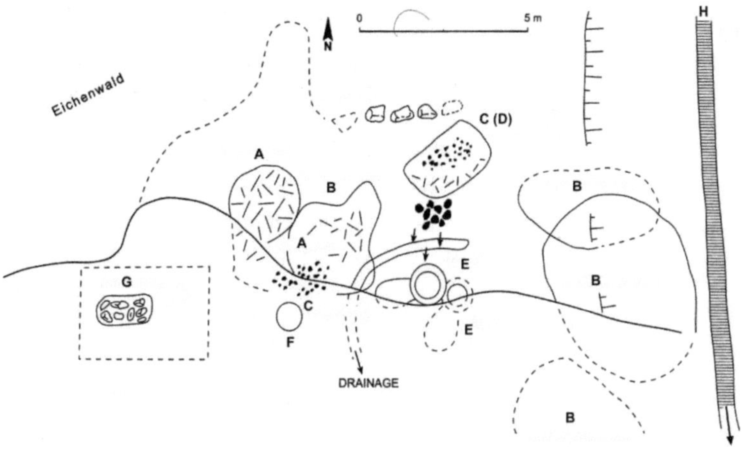

Fig. 13.4 Typical medieval site of metallurgy in the NW part of Hesse Province, in the Dietzhölze Valley (**a** charcoal storage, **b** slag pile, **c** ore crushing, **d** ore roasting, **e** ore-smelting furnace, **f** pit, **g** shelter, **h** stream) (Willms 1998)

In many countries of Western Europe, during the Middle Ages and the Early Modern Age, in low-lying, often waterlogged areas, certain industrial objects (primarily windmills) were placed upon *artificial mounds* (Plate 13.3).

Plate 13.3 Artificial mounds of windmills in Northern Europe (Sources: http://judy-tom.com/images/Windmill.jpg and http://citymitten.files.wordpress.com/2009/08/dscn6850.jpg)

The morphology of *hollows*, cellars and other *passages* for various purposes (defence, storage, underground transportation) are also the subjects of anthropogenic geomorphology. In the Province of Cappadocia of Turkey, proper 'underground towns' were carved in rocks. Various cellar systems for winery and other storage purposes are related to various rock types and widely used in Hungary. They had to be surveyed for engineering geology in several towns of Hungary (Pécs, Eger, Miskolc) as collapses were frequent. Although a comprehensive study of landform types (cellar morphology) related to the various types of rocks has not been conducted, in many cases such research provided results that could be generalised. For example, in the case of Eger, the stability conditions and ground plans of cellars carved into various rocks are well known (Table 13.1, Fig. 13.5, Kleb 1978). Within the world heritage bid of 2005, Eger planned to use its cellar network for the purposes of tourism.

The cellars in the town of Pécs also caused serious problems. The density of cellars rivals that of surface building (Balázs and Kraft 1998). These cellars are

Table 13.1 The types of cellars in Eger according to rock material and ground plan (after Kleb 1978)

Rock material	1. Tuff-cellar	2. Sandstone or marl cellar	3. Cellars carved into travertine	4. 'Gravel cellar'
Ground plan	Hole cellars with pillar strips of parallel	Hole cellars with pillar strips of parallel	Irregular ground plan and walls	Hollows of some metres across
	Finger-like branching	Finger-like branching	Irregular halls	

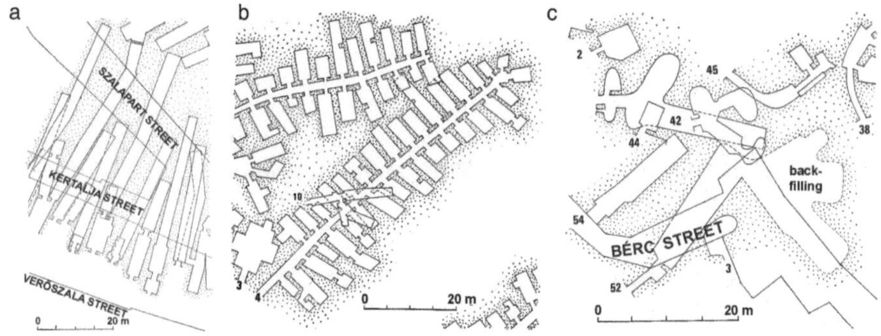

Fig. 13.5 Types of the cellar in Eger by ground plan: (**a**) tuff-cellar; (**b**) sandstone, marl cellar; (**c**) cellars carved into travertine (Kleb 1978)

one-, two- or multi-storied. Also, linkage between these cellars of various owners often took place by emergency corridors to which inhabitants were forced by bomb attacks during World War II. The main use, however, was for wine storage as in other wine-growing areas of Hungary (the Tokaj-Hegyalja, Bükkalja and Mátraalja regions).

At 'Farkasmály' near Gyöngyös, a coherent system of cellars with an area of more than 1,000 m^2 was developed, providing shelter to the population of the surrounding areas during World War II (Plate 13.4). Subsidence, collapse and slumping

Plate 13.4 The row of cellars at Farkasmály near Gyöngyös; the size of the underground passage is indicated by that of the barrels (Dávid 2005)

took place when cellars caved in causing severe damage in buildings and roads in many places.

13.3 The Impacts of Transportation on the Surface in the Modern Age

13.3.1 Construction of Transport Network: Direct Impacts

Historically, four periods of geomorphological interventions are distinguished: primeval, Roman, post-Roman and modern. Since the Modern age, road construction resulted in increasingly more profound changes on the Earth's surface (Sherlock 1922). The growing demands of passenger and freight transport since the Industrial Revolution also led to landscape transformation. The rapid economic development started with the Great Explorations and colonisation resulted in significant advancement both in *land* and *inland-water transportation*. On land, high-quality public roads, tunnels crossing the mountains as well as channels and chain bridges were constructed; however, this was also the time when the first rail tracks were established. Wooden rail tracks were in use as early as the 16th century in present-day Germany, but cast iron rails for horse-cars only appeared in England during the second half of the 18th century. The next important step was the invention of the steam engine. Steam locomotives appeared in the early 1800s, leading to a rapid progress in rail transportation. The total length of the world's railway lines, between 1840 and 1880, increased from 8,000 km to 360,000 km along with rail transportation commencing on all continents. The extent of the routes of rail transport and the network of navigable channels are indicated in Fig. 13.6

The post-World War I era saw the beginning of a rapid development in road transportation that was due to the spreading of vehicles. Increased motorcycle, car and bus traffic, however, required the modernisation of the public road network. The first motorways ('highways') with 2 + 2 tracks, separating traffic and without level crossings, were built in the USA in the 1920s; however, they became apparent in Germany only 10 years later. The first motorways meeting both the demands of the modern times and conforming to the present-day definition were built in Germany and the country's network, by the early 1940s, exceeded 2100 km. Apart from Germany, motorways were only constructed in the Netherlands before World War II. Similar but lower-quality roads were also built in Italy between the second half of the 1920s and World War II, for a length of 500 km.

Following World War II to the mid-1970s, 70,000 km motorways were constructed in the USA, financed by the federal government. West Germany, within the framework of a 4-year federal road construction plan launched in 1957, developed its motorway system making up one-fourth of the European network. In France, government policy preferred, for a long time, railways and started to focus on – except for urban motorway sections – the construction of motorways in concessions since the 1970s.

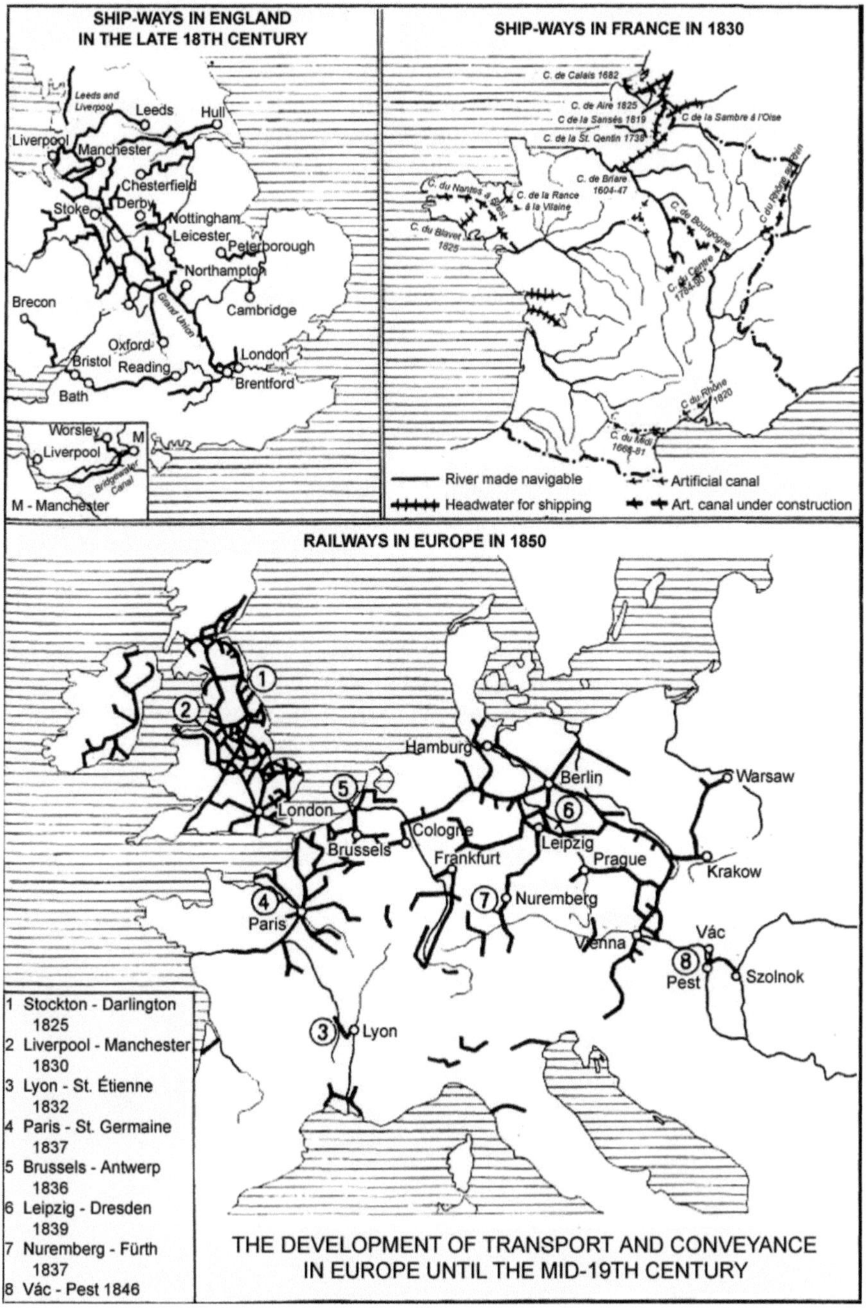

Fig. 13.6 The network of rail transportation and navigable channels in Europe until the mid-19th century (Urbán 1972)

Public road development in Hungary was rather slow. In the 1930s, only 10% of the country's roads had a firm surface (Kovács 2002). During the 1940s, asphalt paving on macadam roads began. During World War II, along with railways, most of the public roads and bridges were also destroyed or damaged. More significant developments and asphalt surfacing took place only from the 1960s due to large-scale motorisation. Along with the closing down of low-traffic railway lines, road constructions and restorations of contemporary, particularly low technical quality, roads also took place. Motorway constructions in Hungary, along with decentralisation initiatives, also started then (Erdősi 1969). In the 1970s, roads were paved with asphalt and main roads were modernised and widened (Szentpéteri 1999). The present programmes of road network development affect vast areas in the country (Fig. 13.7).

In addition to the benefits detailed above, the dramatic impact of motorways on their environment should not be neglected. First of all, they have a vast area demand. About 67 ha per km are affected by the local microclimate of motorways and the exhaust gases of vehicles, thus agricultural products grown next to motorways do not meet strict standards. A 1–2 km strip of the motorways suffers ecological damage that is already apparent during the construction works as vast amounts of material are transported to the construction site. This means a great pressure on the already existing road network.

Within the systems of road transportation, spectacular motorway constructions are presently taking place in Hungary. Both during their construction and operation, motorways have a rather complex impact on their environment. In addition to area demand, the large-scale earthworks severely damage both the abiotic and

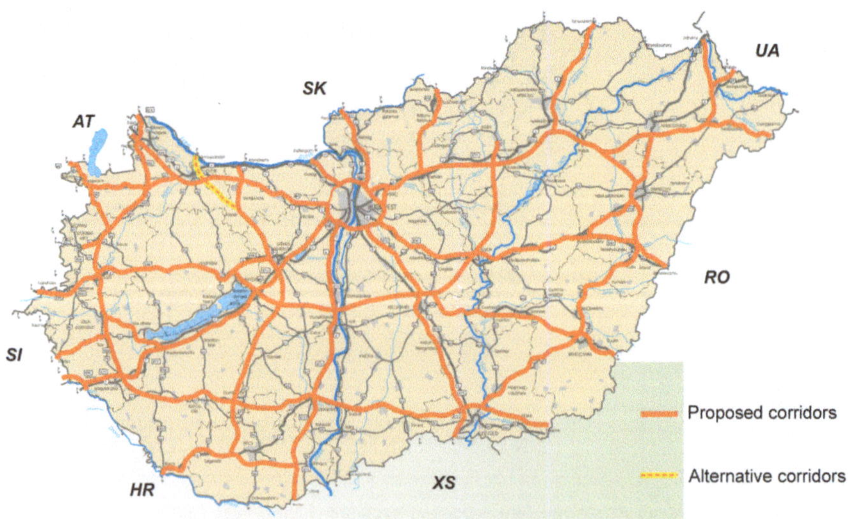

Fig. 13.7 Proposed high-speed road network in Hungary until 2030 (Source: Institute for Transport Sciences Non-profit Ltd. http://www.kti.hu)

biotic environments. Motorway construction is an activity involving a compulsory environmental impact assessment during which landscape scars also have to be studied in detail. The rate of the proposed interventions and the resulting landforms as well as how the motorway is best adjusted to the landscape have to be assessed.

The construction of a 1 km motorway section, 28 m wide separate, 2 + 2 lanes (with 2 emergency lanes) requires 8 ha of land. During the construction pits, quarries have to be opened and subsidiary roads built for the exploitation of building materials. Creating noise barriers, game crossings and other structures for environmental and nature conservation purposes may lead to significant alterations of the surface.

During the construction, significant soil erosion may occur. During operation, pollution caused by oil, gas and heavy metal emissions and tyre wear of vehicles damage surface and underground waters directly or indirectly and may contribute to soil degradation.

In Hungary, the geomorphic impact of transport network was studied by Erdősi (1987) in the environs of the Mecsek Mountains (Figs. 13.8 and 13.9).

For landscape aesthetics, roads are enduring artificial landscape elements. This is represented by, on the one hand, the ecological effects already discussed and, on the other, by their mostly predominant presence in the landscape. A road has to fit in with the basic character of the surrounding landscape in a landscape ecological, functional and aesthetic context (Csemez and Csima 1989). The sight of construction works is also important: tunnel and subways are hardly visible, while flyovers, bridges and embankments several metres high are marked elements of the landscape.

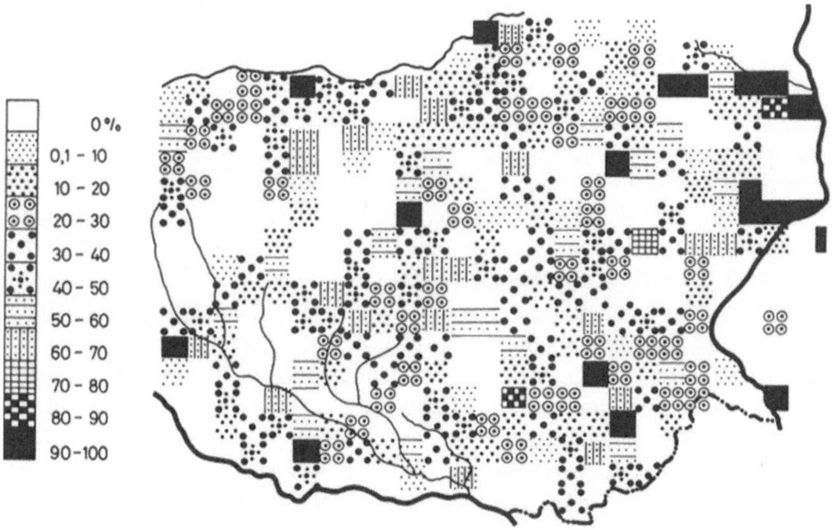

Fig. 13.8 The relative proportion of the length of embankments and cuts to the total length of road and railway lines (in percentage) (Erdősi 1987)

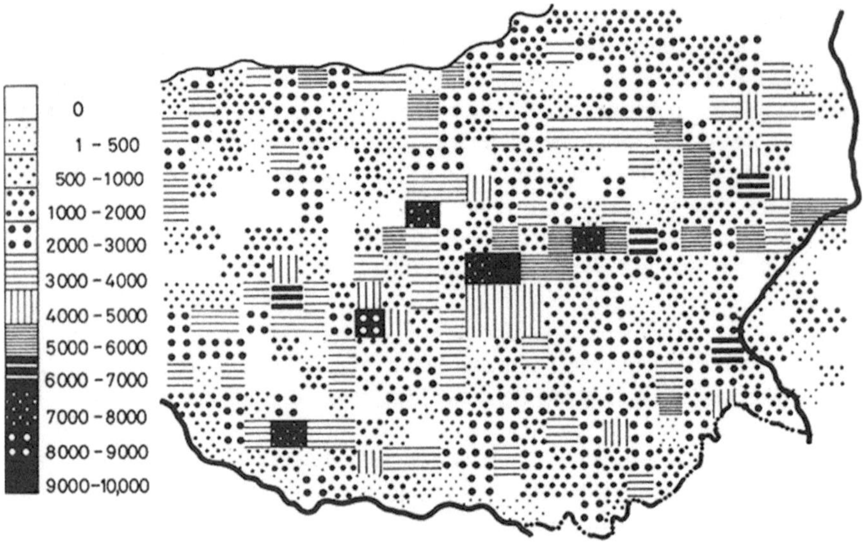

Fig. 13.9 The spatial distribution of surface modification through the construction of transport routes, the relative extent of earthworks in m^3 per km^2 (Erdősi 1987)

Tunnels are also part of railway and road transportation networks. Tunnel building is said to have begun ca. 4160 years ago, when Queen Semiramis, famous for her terraced gardens, one of the seven wonders of the ancient world, had a tunnel 1 km long and of 16 km^2 cross-section area built under the River Euphrates to link her castle and the Temple of Jupiter. To our knowledge, the Egyptians also built tunnels. A remarkable construction in Jerusalem, built 2700 years ago, conducted spring water to the town. The ancestor of road tunnels crossed a hill between Naples and Pozzuoli 2000 years ago. Under the reign of Emperor Augustus a tunnel of 900 m length and 7.5 m width was completed. During Medieval times, achievements of foundation engineering were represented by the ramparts and tunnel systems of fortresses and mine adits. A significant development in tunnel construction was initiated only in the 17th century Europe, especially in France. It was the engineer Francois Andreossy, who in 1666 first opened a tunnel by explosion into rocks in the Province of Languedoc, Southern France, to the navigation canal known as Canal du Midi, 20 m in width and 240 km in length, linking the Atlantic Ocean and the Mediterranean Sea. The growing trade required a developed road network instead of narrow roads and passes inaccessible for long periods of the year. The construction of a tunnel under the English Channel was a dream recurring from the early 19th century. Between 1807 and 1842, the 1,100 m long tunnel under the River Thames was built in London. The first railway tunnel near the town of St. Etienne in France was constructed for horse-trams later replaced by steam locomotives. The first real mountain railway was accomplished in Austria in 1854. The 41 km-long Semmering railway line peaks at a height 899 m above sea level in the 1,428 m long Semmering Tunnel. On its route, the railway passes further 14 tunnels. In urban transport, a turning point came on 10th January 1863, when the first section of the Metropolitan Line of the London Underground between Paddington and Farringdon was opened. The rapidly expanding high-speed special surface and underground railway lines was challenging from other aspects. Here, damage to the built environment, through, e.g. subsidence has to be avoided, special attention has to be paid to poor soil conditions (compared to mountains); neither urban transport nor city life can be disturbed and the environment is of primary importance. Today, several hundreds of kilometres of tunnels are being constructed each year to conduct water as well as for railways and roads (Table 13.2), urban communal and transportation purposes. Underground lines operate or are being constructed in nearly a hundred cities of the world. Ambitious plans and proposals have been made, among which, the tunnel under the Strait of Gibraltar between Europe and Africa, a railway tunnel between Innsbruck and Italy under the Brenner Pass of 60 km length.

The construction technology of underground structures is determined by several factors. Apart from geological and geomorphologic aspects (soil structure, strength, permeability, groundwater level, etc.), geometry and location (floor depth, earth cover above the floor, etc.) are also important. Also, a decisive factor is the impact the construction technology selected is expected to have on the built and physical environment. Its impact on the human environment in the densely built-up urban environment should not be neglected either. The appropriate construction technology should be designated to match subsidence hazard and other risks.

As a result of *air transport,* appearing from the early 20th century, the number of airports has been increasing, and the associated planation activities affect large areas. It is difficult to estimate the total number of airports and runways in the world.

Table 13.2 The world's longest railway tunnels in 2008 (Source: http://www.robl.w1.com/Transport/tunnel.htm)

Railway tunnel	Country	Length (m)	Year of opening
Seikan	Japan	53,841	1988
Euro-Tunnel	France–United Kingdom	50,500	1994
Lötschberg Base	Switzerland	37,577	2007
Guadarrama	Spain	28,377	2007
Taihang	China	27,848	2007
Iwate-Ichinohe	Japan	25,810	2002
Shimizu III.	Japan	22,300	1982
Wushaoling	China	20,050	2006
Simplon II.	Italy–Switzerland	19,824	1922
Simplon I	Italy–Switzerland	19,799	1906
Shin-Kamnon	Japan	18,600	1974
Appennino	Italy	18,507	1934
Qinling	China	18,457	2002
Zhongnanshan	China	18,040	2007
Rokko	Japan	16,214	1972
Furka	Switzerland	15,442	1982
St.Gotthard	Switzerland	15,003	1881

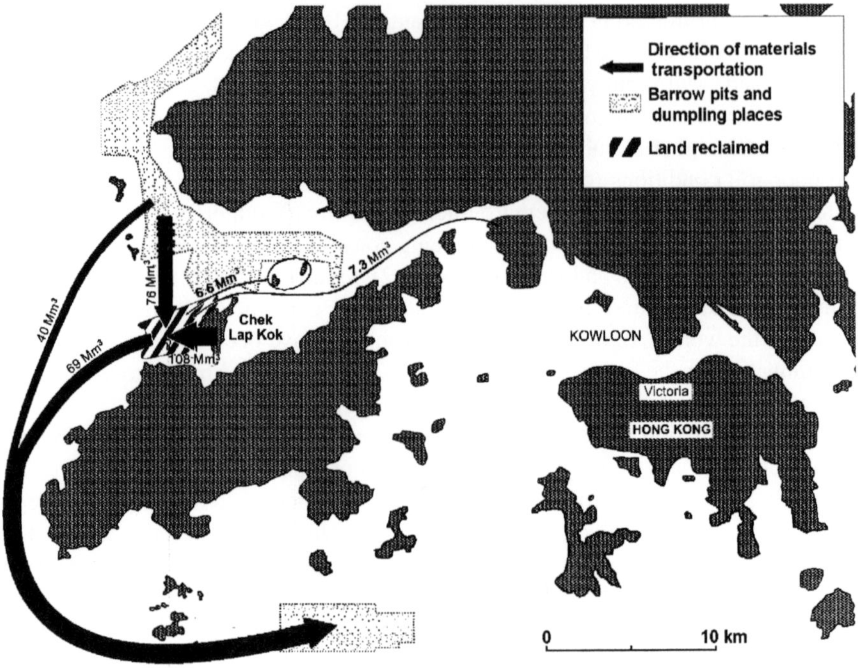

Fig. 13.10 Material removal during the construction of the Chep Lap Kok Airport in Hong Kong (Rózsa 2004)

In the United States alone more than 10,000 of them are found, although airfields with no concrete runway have only limited surface sealing impact. Reclaimed land is increasingly favoured for runway construction. In Hong Kong, a city of limited extension and mountainous area, the airport was built on land filled with construction rubble. The establishment of the new airport on the Isle of Chek Lap Kok during the 1990s involved one of the largest earth removals in history. During this, on the western side of the granite island with an area of 302 ha, 938 ha of new land was reclaimed, and extended the shoreline by approximately 5 km. From the future area of the airports, about 69 million m^3 of silt was obtained from the offshore sea bed; approximately 76 million m^3 of sand was imported here. Further 40 million m^3 silt and alluvial material had to be moved during the works. One hundred and twenty-two million m^3 of rocks were excavated on land (mostly on the island itself), of which the 13 km-long sea-wall surrounding the airport island was built of granite blocks of 1–5 tonnes. Over the levelled surface of the island, 2 m of crushed granite and sand was spread. The material removed during the construction of the Chek Lap Kok Airport in Hong Kong, ca. 307 million m^3, is 18-fold larger than the total amount of construction gravel and sand extracted in Hungary in 2001 (17 million m^3) (Fig. 13.10; Rózsa 2004).

To facilitate *water transportation*, inland waterways (canals) have been built since the end of the 17th century. During their construction, several thousand million cubic metres of earth were directly removed. Prior to the use of steam and diesel engines, low-speed vessels were hauled by man- or horse-power. Consequently, they made minor damage to canal banks. With the increasing speed of vessels, bank erosion has become apparent especially in meanders and necessitates permanent maintenance in order to prevent catastrophic erosion.

Canals shortening sea routes, crossing isthmuses were started to be built since the last third of the 19th century. Among them, the Corinthos, Suez and Panama Canals are the most important. The Corinthos Canal, opened for shipping in 1893, is relatively small: 6.4 km long, 23 m wide and 9 m deep. Crossing hard limestone beds, it has no erosion or landslide hazard. The Suez Canal is located in a rather different environment: it crosses 162 km of sand dunes and saline marshlands. It has an average width of 60 m and a depth of 12 m; during its construction, ca. 110 million m^3 material was removed. In the late 1970s, a new programme was launched to widen and deepen the canal. Since its opening 120 years ago, two geomorphologic factors contribute to the accumulation of sand here: wind action and bank collapses. The latter are caused by waves beyond crossing ships. To achieve a constant floor shape, continuous dredging is required as evidenced by the Middle East wars when maintenance works were ceased. Of the three channels mentioned above, the Panama Canal had the most remarkable geomorphological impact. The 64 km-long Panama Canal was built between 1882 and 1914. It crosses a difficult terrain and demanded 100 m deep cuts. Steep slopes resulted in unstable rock and the hot, humid tropical environment caused landslides so huge that the works had to be stopped. Some claimed that during a single night, 382,000 m^3 material moved downslope in a single mass. The removal of approximately 375 million m^3 of earth was necessary; this is nearly fourfold of the total annual bed load of the Mississippi River (Nir 1983). Also, the fact this vast material transport is concentrated on a rather small area should not be neglected. The widening of the Panama Canal has begun and will be completed by 2012.

13.3.2 Indirect Impacts

Paved roads seal the surface and prevent conventional erosion from acting. However, they induce other geomorphic processes. The indirect impacts of road constructions are also extremely important (Nir 1983). They are the following:

- impacts on the impermeable layer,
- changes in runoff,
- the ‘waterfall effect’,
- environmental impacts under permafrost conditions.

13.3.2.1 Cutting Through Impermeable Layers

Public Roads

In road-cuts the hydrological equilibrium is upset, the cut slope is exposed to landslides, as well as the likelihood of rockfalls and creep increases. Some landslides of this kind occurred along the slopes of the Jerusalem–Tel Aviv motorway, crossing mountains, in the 1960s. The limestone slopes above the impermeable marl began to slide and large-scale mitigation works had to be made. In this region, only slopes with an angle less than 25% remained stable (Fig. 13.11; Nir 1983).

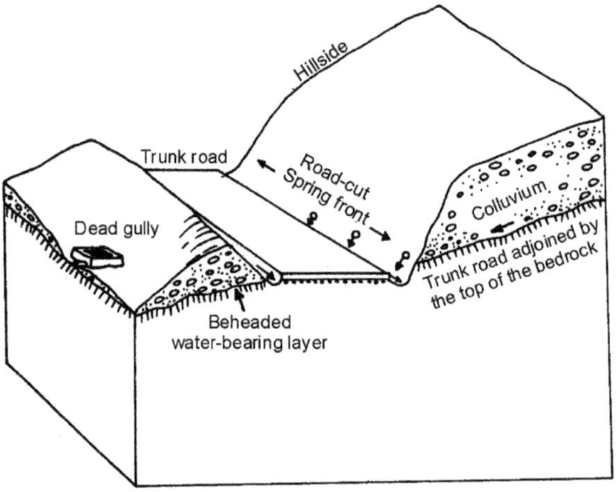

Fig. 13.11 The impact of road-cut on groundwater supply. The risk of landslides is increased along the spring line (Nir 1983)

Railways

From the point of view of earthworks, the main difference between roads and railway tracks is that the latter have a lower potential maximum gradient. Therefore, railway cuts are deeper. In practice, the excavated material is used in railway embankments in the same amounts. On terrains of more complex topography, tunnels are constructed. Rail tracks have a limited impact on erosion as the loose crushed stone cover concentrates runoff to a lesser degree – provided bridges and culverts are built in adequate numbers. In cuts, however, erosion control techniques have to be applied. The impact of accelerated erosion there is dangerous – especially in semiarid areas. Slides due to railway constructions are larger in scale than those occurring during road construction. As roads overcome steep slopes by curves and serpentines, local cuts expose bare surfaces; however, the total area of cuts can be remarkable. As railway tracks demand gentler slopes, cuts expose larger surfaces, although the number of cuts (per kilometre) may be less. As erosion is, among other factors, dependent on the length of the slope exposed, in the case of impact areas of the same size the likelihood of erosion is higher for railway lines than for roads. In India, along the 170 km-long Assam–Bengal railway line, intense erosion caused so severe a damage in 1915 that restoration required 2 years without traffic. Another good example is the Folkestone–Dover railway that suffered severe damage in 1915 from a landslide jointly triggered by the water beneath the train and formerly retained from the slope (Nir 1983). Landslide hazard represents a nearly permanent problem around Abaliget, along the section of the Budapest–Pécs railway line crossing the Mecsek Mountains (Szabó 1993).

13.3.2.2 Changes in Runoff

Roads

Paved roads collect 90–95% of rainwater runoff and usually release a remarkable amount of water within a short period of time through localised culverts. Erosion caused by flash floods is unavoidable as the available natural river beds were mostly formed by low-energy runoff. As a result, deep channels can form. To avoid badland formation, a number of culverts have to be built to distribute runoff among them (Nir 1983).

Airfields and Airports

It is the vast extensive paved surface of large airports that contributes to runoff concentration (as discussed above). No damage caused by erosion or sedimentation is known for major airports: the need for channelisation had obviously been recognised (Nir 1983).

13.3.2.3 The 'Waterfall Effect'

Most of the erosion along roads takes place on the boundary between paved and unpaved surfaces, where a gap soon develops. To the initial energy of runoff from the paved surface, further energy is added by gravity deriving from relief between the two surfaces. Turbulence triggers a so-called *'waterfall effect'* leading to the erosion of the road embankment under the cover. In case this process unfolds unhindered, the pavement may be cut through in its entire width (Fig. 13.12; Nir 1983).

Pipelines

The network of pipelines is growing constantly both in the developed and developing countries. Compared to other modes of surface transportation, pipelines are often regarded as a moderate intervention into the environment, and when buried cause no more disturbance. The alignment of pipelines being designed to be the shortest possible way, regardless to the topography. Consequently, their impact may be more severe than that of railways. Considerable erosion hazard occurs in two types of environment.

One of them is *arid regions* where no natural vegetation cover protects the surface against wind erosion or gully formation. The other is *arctic regions* with permafrost. Constructions under arctic conditions are presented on the example of the planning of the *Alaska Pipeline*. The relatively warm liquid transported keeps the surface in a thawed condition continuously in the environs of the pipeline. Under circumstances in Alaska, thawing penetrates, in 2 years, to a depth of 4–6 mm (in 9 years to 9 mm), causing a subsidence to a degree that would break the pipelines. Therefore,

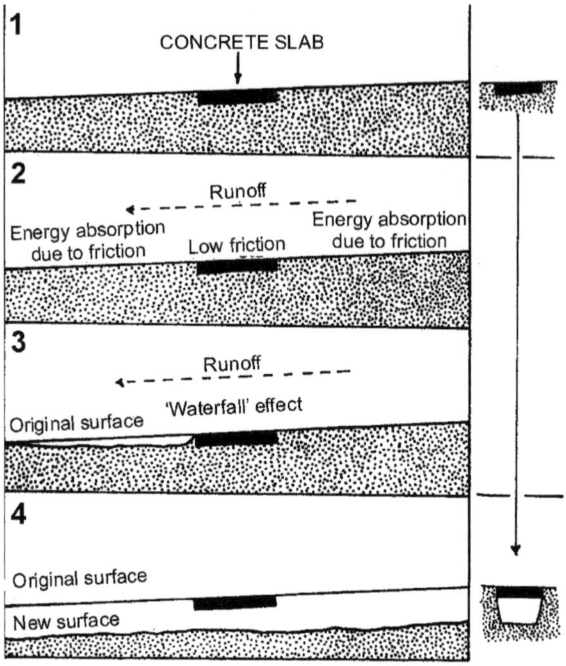

Fig. 13.12 'Waterfall effect' in a ditch lined with concrete slabs to regulate flow (1: initial surface, 2: impact of slab on the frictional energy, 3: the 'waterfall effect', 4: the final stage: incision below the slab (Nir 1983)

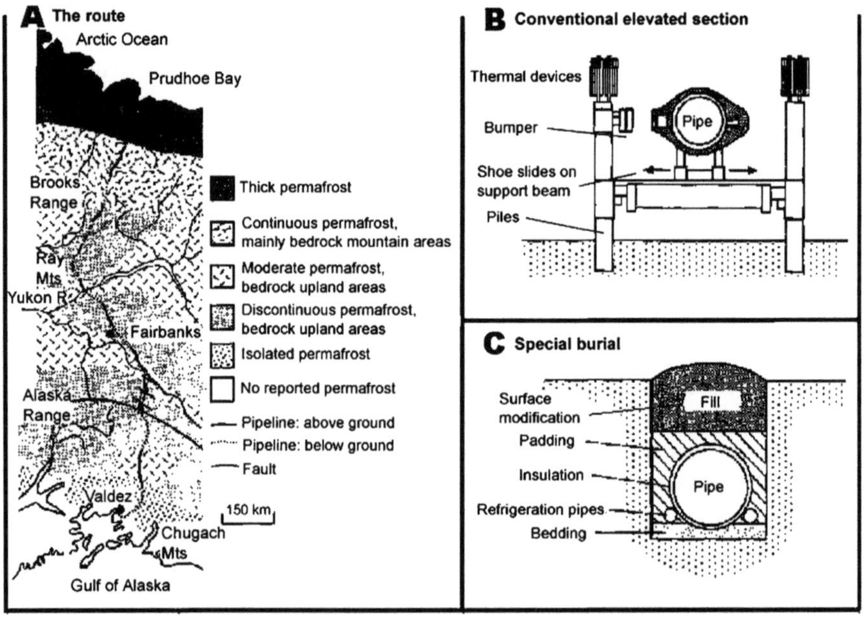

Fig. 13.13 The physical environment and technological solutions of oil pipelines in Alaska (after (Bennett and Doyle 1999, modified by Dávid and Baros in Szabó and Dávid 2006)

the pipeline was constructed above-ground on posts at 30 m distance from each other and the base of these posts was placed in a depth not affected by seasonal thaw. Consequently, the installed pipeline will not influence the equilibrium of permafrost (Nir 1983). There are solutions to this problem like applying a special insulation to the underground pipelines; here, the surface landform (earth embankment) of accumulation origin influences geomorphic processes (Fig. 13.13; Bennett and Doyle 1999).

Compared to any other way of surface transport, pipelines have another characteristic: they are almost entirely *independent of topography*. These routes are the shortest possible, across hills and valleys if made possible by technology and if the transportation of the liquid (i.e. pressure) can be maintained economically. For this very reason, the erosion impact of scars left after laying such pipelines is more intense than in the case of road and railway construction. Scars formed during the construction of pipelines crossing the relief along a straight line, promotes wind erosion – especially in arid regions.

13.3.3 Modern Industrial Development

The suddenly increased industrial production since the Industrial Revolution, also involved significant changes in topography. On the one hand, there has been an increase in the extent of areas affected and, on the other, the intensity of such interventions, and consequently their impact, has become higher. The rate of denudation resultant from constructions and mining is two- to fourfold higher compared to that of natural processes. Some authors claim the presence of a global geomorphological change due to the industrial and mining surface transformation and removal of materials (Rivas et al. 2006).

13.3.4 Sludge Reservoirs

The disposal and neutralisation of waste originated during ore processing, i.e. sludge take place in *sludge reservoirs*. In order to insulate the storage area, it is covered by artificial, multi-layered coating. Various methods are applied to install surplus capacities to accommodate rainstorm water and major surface runoff: watercourses conducted around sludge reservoirs and treating seepage from the sludge reservoir prior to releasing them into surface waters. During the reclamation of sludge reservoirs, occasional major geomorphological interventions are designed to prevent the spreading of contaminations released from them.

In Hungary, during the *mining of uranium ore* for almost four decades, two sludge reservoirs with a total area of 165 ha were established, where ca. 20.3 million tonnes of solid substance (with an average uranium content of 67.87 g/t) and approximately 32 million m^3 of technological solutes were disposed (See Case Study 10.6). These sludge reservoirs are located above two aquifers – source areas important for the drinking water supply of the city of Pécs and partly the surrounding settlements.

Reclamation has been completed recently: following the stabilisation and contouring of the sludge core and its multi-layered covering, re-vegetation efforts are only taking place today (Benkovics 2006).

13.3.5 Slag Cones and Fly-ash Reservoirs

The traces of *metallurgy* are present as *slag cones*, similar to waste heaps of coal mines, alter the topography of disposal sites to a significant degree. The porous material of unreclaimed and unvegetated piles, when affected by rainwater, will be rapidly saturated, causing mass movements on clays with low stability. Toxic materials can be spread by these processes. Mitigation can be achieved by biological land reclamation.

The accumulation of *fly ash* generated in waste combustion sites has similar impacts.

The management of these two problems is also important as the dispersed material (e.g. ignition slag) can contain *toxic* heavy metals, dioxine compounds and other toxic materials in large quantities.

13.3.6 Cooling Ponds

Most of the water used by the industry is *cooling water*. Until past decades, released hot water and used thermal water were not considered to be of deteriorated quality, as the resultant changes are not physical or chemical in character. Recently *heat pollution* has become one of the potentially most harmful type of water pollution (relatively higher water temperature compared to its environment promotes organic matter generation, the proliferation of hydrophytes, upsetting biological equilibrium, irrigation water with high salt and sodium content results in soil alkalisation, etc.).

Power plants, especially nuclear plants, continuously grow in size and the cooling of warm waste water is increasingly difficult to solve.

The heat pollution of surface waters has become an acute problem due to requirements of water recycling and the temperature range modified due to storage and, as a consequence of the high cooling water demand of nuclear power plants planned and operating today, this heat pollution shows a growing tendency worldwide.

Emplacement of thermal water cooled during use represents a similar problem: it is usually drained to public sewers, lateral ditches, occasionally to lakes or reservoirs. Thermal waters, during their use, usually cool down to a temperature that no additional treatment is required. In many cases, however, thermal water utilisation and drainage systems are constructed by having a cooling pond interposed, to provide thermal water exploited from the well (e.g. at system breakdown) to be cooled under 40°C.

Vegetation proliferation resulting from heat pollution will eventually lead to the eutrophication of ponds. The character of geomorphic processes typical for abandoned cooling ponds is determined by the type of *after-use*. A typical example is use as angling ponds where, among others, erosion caused by trapping is predominant.

13.3.7 Industrial Parks, Shopping Centres

Industrial parks, logistic centres, shopping centres and sometimes residential parks, appeared following the change of regime in Hungary, and their increase in numbers ever since (Fig. 13.14), have remarkable impacts on the environment. Construction works (landscaping, deep foundation, drainage, changing runoff, etc.) remove large amounts of earth. Fulfilling the building material demands of construction require intensified mining activities.

Fig. 13.14 Industrial parks in Hungary (as in July 2008) (Source: Institute for Transport Sciences Non-profit Ltd. http://www.kti.hu)

The geomorphologic impacts of access road networks leading to such projects were discussed under the impacts of transportation.

This process is unlikely to slow down in the future. In many cases municipalities are forced, in order to improve their limited financial capabilities and to maintain settlement infrastructure, to ensure that already existing facilities attract further ones (Molnár 2006).

13.3.8 Environmental Impacts of Industrial Waste Disposal

Waste disposal, primarily of *solid communal and industrial waste*, plays an increasing part in the transformation of topographic conditions. The amount (and composition) of solid waste of settlements is dependent on the level of urbanisation, economic development, life quality and life-styles; however, differences occur between rural settlements and cities, or even within urban districts. In Hungary, since 1990, an annual amount of 4–5 million tonnes of solid communal waste has been continuously generated, with its relative percentage – due to the lowering amount of industrial and agricultural waste – increasing. As this type represented about 7% of all wastes in 1990, this rate by 1995 increased to 9% and by 1999 to more than 10% (Table 13.3).

Table 13.3 The amounts of various waste types generated in Hungary, collected and transported in an organised way (in million tonnes) (Rózsa 2004)

Waste type	1990	1995	1999
Agriculture and food industry, non-hazardous	13.0	4.0	5.0
Industry and other production, non-hazardous	34.6	27.1	23.2
Solid communal	*4.9*	*4.5*	*4.5*
Communal fluid	11.7	9.6	6.3
Communal sewage sludge	0.3	0.4	0.5
Hazardous	4.5	3.4	3.7
Total	69.0	49.0	43.2

In the late 1990s, Hungary's ca. 2700 *disposal sites* for solid communal waste (of which about 400–500 were operated legally) occupied an area of 75–80 km^2. Waste disposal was, for long decades, typically resolved by landfill in open pits, quarries, waterlogged areas and other excavation landforms around settlements. The establishment of further sites rather involves the shaping of smaller hummocks or positive landforms. Some larger waste disposal sites are of several square kilometres in area, and their final relative height following compaction can even reach 10–20 m (Rózsa 2004). Plate 13.5 shows the growth of (tyre) disposal sites of solid waste.

Plate 13.5 Tyre disposal sites (Sources: http://www.ecodepot.co.nz/photos/lTyres.jpg and http://product-image.tradeindia.com/00228701/b/0/Waste-Tyres.jpg)

13.4 Summary

The most common positive, negative or levelled landforms resulting from transportation or industrial activities are summarised in Table 13.4.

Table 13.4 Landforms due to transportation and industrial activities (by Dávid in Szabó and Dávid 2006)

Landforms Resultant From Transportation Activities										
Overland							Air		Piped transportation	
Road		Railway		Water			Pipe		Electric	
+	–	+	–	+	–	+ *and* –	+	–	+	–
Embankment	Sunken road	Embankment	Cut	Embankment	Channel	Airport	Embankment	Ditch	Embankment	–
Dam	Cut	Dam	Tunnel	Dam	Ditch	Airport,	Mound	Ditch	Mound	–
	tunnel					infra–structure				
Landforms Resultant From Industrial Activities										
+			–				*Planation*			
Mound of windmill			Cooling pond				Industrial park surface			
Slag cones and fly-ash reservoirs			Sludge reservoir				Solid waste disposal site			
Solid waste disposal site			Waste disposal basin				Industrial area			

References

Ádám L (1964) A Szekszárdi-dombvidék kialakulása és morfológiája (Development and Morphology of the Szekszárd Hills). Akadémiai Kiadó, Budapest

Balázs F, Kraft J (1998) Pécs város településének mérnökgeológiai vonatkozásai (Engineering Geological Aspects of the Town of Pécs). Pécs

Benkovics I, Bánik J, Berta Zs, Csővári M, Csicsák J, Földing G, Várhegyi A (2006) A magyar uránipar felszámolásának és rekultivációjának programja (Liquidation and recultivation programmes of the uranium industry in Hungary). A VIII. Bányászati, Kohászati és Földtani Konferencia (Sepsiszentgyörgy, 2006. április 6–9.) előadáskötete (Papers of the 7th Conference on Mining, Metallurgy and Geology, 6–9 April, 2006). Erdélyi Magyar Műszaki Tudományos Társaság

Bennett MR, Doyle P (1999) Environmental Geology. John Wiley and Sons, Chichester

BorsyZ (ed.) (1993) Általános természetföldrajz (Physical Geography). Nemzeti Tankönyvkiadó, Budapest

Csemez A, Csima P (1989) Tájrendezés (Landscaping) II – Kertészeti és Élelmiszeripari Egyetem Kertészeti Kar, Budapest

Denecke D (1969) Methodische Untersuchungen zur historisch-geographischen Wegeforschung im Raum zwischen Solling und Harz. Göttinger Geographische Abhandlungen, Göttingen

Endrédi L (2001) A Tolnai-Sárköz és a Szekszárdi-dombság természeti környezeti értékei (Physical environmental values of the Tolnai-Sárköz and the Szekszár Hills). In: Endrédi L (ed.), Szekszárd vidékének természeti és kulturális környezeti értékei. Pécsi Tudományegyetem Illyés Gyula Főiskolai Kara, Szekszárd. http://gutenberg.igyfk.pte.hu/edok/szekszard/kornyezet.htm

Erdősi F (1969) Az antropogén geomorfológia mint új földrajzi tudományág (Anthropogenic geomorphology as a new discipline of geography). Földr Közl 17: 11–26

Erdősi F (1987) A társadalom hatása a felszínre, a vizekre és az éghajlatra a Mecsek tágabb környezetében (The Impact of Society on the Surface, Water and Climate in the Broader Environment of the Mecsek Mountains). Akadémiai Kiadó, Budapest

Gönczy S, Szalay K (2004) Geomorfológiai fogalomgyűjtemény (An Anthology of Geomorphology). Kárpátaljai Magyar Pedagógusszövetség Tankönyv- és Taneszköztanácsa, Beregszász. http://mek.oszk.hu/02900/02911/02911.htm

Ilyés Z (2004) A montániparok és az erdészet vízgazdálkodása a Kárpátokban (Water management of the mining industries and the silviculture in the Carpathians). In: Füleky Gy (ed.), A táj változásai a Kárpát-medencében. Víz a tájban. Környezetkímélő Agrokémiáért Alapítvány – Szent István Egyetem, Gödöllő

Kleb B (1978) Eger múltja a jelenben (The Past of Eger in Its Present). Budapest

Klischat S (1996) Alle Strassen führen nach Rom – die Via Appia und andere Römerstrassen. http://www.klischat.net/onlnepub/referate/rom/rom.htm

Kovács F (2002) Közlekedéstan (Transportation Studies) e-jegyzet. http://zeus.szif.hu/ejegyzet/ejegyzet/kozlekedestan/10whtml.htm

Molnár JL (2006) Városszéli bevásárlóközpontok tájformáló hatása (Landscape-forming Impact of Urban-Fringe Shopping Centres). A II. Magyar Tájökológiai Konferencia Abstract kötet, Debrecen

Nir D (1983) Man, A Geomorphological Agent. Keter Publishing House, Jerusalem, Boston, London

Rathjens C (1979) Die Formung der Erdoberfläche unter dem Einfluß des Menschen. Grundzüge der Anthropogenetischen Geomorphologie. B. G. Teubner, Stuttgart

Rivas V, Cendrero A, Hurtado M, Cabral M, Giménez J, Forte L, del Rio L, Cantú M, Becker A (2006) Geomorphic consequences of urban development and mining activities; an analysis of study areas in Spain and Argentina. Geomorph 73: 185–206

Rózsa P (2004) Város és környezet (The city and the environment). Debrecen

Rübl J (1959) A Pilisvörösvári-medence geomorfológiája (Geomorphology of the Pilisvörösvári Basin). Doktori disszertáció. http://vorosvar.freeweb.hu/Szak/Rubl.htm
Sherlock RL (1922) Man as Geological Agent. Witherby, London
Szabó J (1993) A társadalom hatása a földfelszínre – antropogén geomorfológia (The impact of society on the Earth's surface – Anthropogenic geomorphology). In: Borsy Z (ed.) Általános természetföldrajz. Nemzeti Tankönyvkiadó, Budapest
Szabó J, Dávid L (eds) (2006) Antropogén geomorfológia (Anthropogenic Geopmorphology) University notes. Kossuth Egyetemi Kiadó, Debrecen (the Hungarian version of this book)
Szentpéteri É (1999) A magyar közúthálózat állapota (State of the public roads in Hungary). Lélegzet 9 (7–8)
Szilágyi M (ed.) (2001) Gazdálkodás (Farming). In: Magyar Néprajz nyolc kötetben. MTA Néprajzi kutatóintézet, Budapest. http://mek.tlkk.org/02100/02152/html/02/380.html
Urbán A (1972) Újkori egyetemes történet 1789–1918 (History of the Modern Age 1789–1918). Térképvázlat-gyűjtemény. Tankönyvkiadó, Budapest
Willms C (1998) Frühindustrielle Landschaften nördlich von Dillenburg und im Märkischen Sauerland. Sielungsforschung. Archäologie-Geschichte-Geographie 16: 339–362

Website Sources

http://www.wikipedia.de
http://www.kvvm.hu/szakmai/karmentes/kiadvanyok/fav/tvkm/tvkm05.htm
http://judy-tom.com/images/Windmill.jpg
http://citymitten.files.wordpress.com/2009/08/dscn6850.jpg
http://www.robl.w1.com/Transport/tunnel.htm)
http://www.ecodepot.co.nz/photos/lTyres.jpg
http://product-image.tradeindia.com/00228701/b/0/Waste-Tyres.jpg
http://www.parkinfo.hu
http://gtk.wigner.bme.hu/jegyzet/jegyzetm
http://www.kti.hu

Chapter 14
Military Activities: Warfare and Defence

Zoltán Ilyés

Abstract Human landforms created for the purposes of defence include defence lines, fortifications, etc. and have preventive or protective roles and were mostly built in peacetime in a planned way. They are based on strategic concepts of war conduct, reflecting contemporary technical standards. On the other hand, wars also bear significant direct geomorphologic impacts: warfare actions often result in degraded areas. The impacts of bombs, mines and grenades produce negative and positive landforms of various sizes, reflecting the scale, destroying power and ballistic features of the explosive. Their persistence and visibility depend on the physical conditions (soil and bedrock, climate, vegetation, relief and type and intensity of land use). Examples are provided for historical defence structures (earthworks, defence walls, medieval ramparts and fortresses), landscape transformations during World Wars I and II as well as the geomorphic impacts of modern wars and nuclear tests.

Keywords Defence constructions · Fortresses · Trenches · Battlefields · Nuclear weapons

14.1 Classification of Landforms of Warfare Origin

Anthropogenic landforms deriving from warfare and defence are usually classified into two major groups according to their formation, development and function. On the one hand, facilities for the protection of strategic points, villages, towns and regions (ramparts, fortification lines, earth mottes, military roads and air bases, etc.) are distinguished as man-made landforms for the purpose of defence. Defence lines, fortifications, etc. have preventive or protective roles and were mostly built in

Z. Ilyés (✉)
Department for Cultural and Visual Anthropology, University of Miskolc, 3515 Miskolc-Egyetemváros, Hungary
e-mail: ilyesz@freemail.hu

J. Szabó et al. (eds.), *Anthropogenic Geomorphology*,
DOI 10.1007/978-90-481-3058-0_14,

peacetime in a planned way. Their construction was based on systematically defined strategic concepts of war conduct, reflecting contemporary technical standards. On the other hand, degraded areas often result from warfare actions as significant direct geomorphologic impacts. The impacts of bombs, mines and grenades produce smaller or larger negative and positive landforms reflecting the scale, destroying power and ballistic features of the explosive. Their persistence and visibility depend on the physical conditions (soil and bedrock, climate, vegetation, relief and type and intensity of land use). Explosions, especially in high mountains, can trigger various mass movements including rockfalls. Moving large amounts of ammunition and technical tools by human or animal power or by machines for military actions involves a range of semi-natural processes. (According to Mortensen (1954/1955), semi-natural processes spring from conditions previously created by human action and follow natural laws over a long period of time. In this respect, it is similar to natural geomorphic evolution, only triggered and maintained by human impact. A similar interpretation is voiced by Szabó (1993).) As a result of treading, for instance, linear microforms develop; mass movement processes are occasionally triggered in arid or semiarid areas; aeolian action locally intensifies (Nir 1983). During warfare actions, damage is done to certain technical structures such as railway embankments, roads, tunnels, dams, barrages and channels. After clearing away the ruins of war and piling the debris, major rubble heaps were formed all over Europe.

Surface alterations taking place at military training grounds cannot be referred into either of the above groups, as they are undoubtedly planned and localised; however, during practising 'real' but imitated military actions affecting a relatively small area, renew man-made landforms continuously (Fehn 1997).

The distinction and classification of anthropogenic landscape elements of military origin are aggravated by the fact that they are mostly combined with other engineering structures. In medieval times ramparts and early earth mottes were constructed by combining pile-works or earth-filled drag-structures; the construction of rock fortifications also demanded the transport of significant amounts of rocks and earth; ramparts were usually supported by stone and brick-works. In the 20th century, most bunkers and fortress lines were built of reinforced concrete; however, for creating their final form remarkable earthworks were carried out: large amounts of earth were piled up on their top or along their walls, steep slopes were created, tank-traps and moats were lined with concrete. Consequently, to obtain an overall picture on man-made landforms with military-defence functions, research methods and documentation techniques from disciplines like the history of arts and architecture, conservation of historical monuments and archaeology should be applied.

Cultural geography traditionally interpreted landscape as a 'palimpsest' and intended to explore, date and assess the contributions made in various times to the present-day cultural landscape. (A palimpsest is a manuscript from which former writings were scraped out and re-written once or repeatedly. Earlier writings preserved in details and in poor quality can be made visible and read by applying special techniques.) The palimpsest metaphor, in general, neatly refers to the historical evolution and morphogenetic approach to cultural landscapes as well as the reconstruction opportunities of cultural landscape history 'layer by layer' (Hard 1973). The studied historical remnants are preserved in rather diverse functions, dimensions and quality, both in time and space. Anthropogenic geomorphological research significantly contributes to such studies traditionally (Hard 1973; Schlüter 1926). Elements related to defence, border security and military strategy are apparent in a higher density in cultural landscapes, commonly – and sometimes unreasonably – referred into *military, defence or march landscape* types (Erdősi 1969 after Bondarchuk). Such landscapes or landscape sections can be regarded as a kind of a 'military palimpsest'.

Anthropogenic geomorphology can play a major role in the identification and inventory of remnants, severely damaged and hardly recognisable (see aerial archaeology). Another topic covered is the classification of blasted bunkers, ruined defence facilities (often disregarded in academic studies). The sphere of competency of anthropogenic geomorphology also has to be clarified in this respect. A solution may be a regulation introduced in Germany where ruined structures as well as their traces in microtopography are protected as archaeological sites and underground historical monuments (Bodendenkmal).

14.2 The largest Defence Line: The Great Wall of China

The Great Wall of China built for more than 2,000 years is the world's largest anthropogenic structure for its length and mass. It has a length of 7,240 km, ranging from the coasts to the east of Beijing and in the north-west to Gansu Province, sometimes in a number of parallel lines. Some of its sections around Lake Lop-Nor have been recently discovered, and its signal-towers can be traced to Kashgar where it secured the former Silk Road. The main wall section is 2,400 km long. Data on the construction of the first section of the Great Wall of China are available from the period between 1100 and 223 B.C. An important phase of construction began between 221 and 207 B.C., under the rule of the Qin Dynasty when the area had to be protected from nomads attacking from the north – primarily against the Xiongnu tribes. Wall sections surrounding Beijing were erected of burnt brick and ashlar stones during the Ming Dynasty between 1368 and 1644 against the Mongols, following the crests of mountain ridges. The gap between walls was filled up with earth, stones and limestone onto which 4–5 layers of bricks were placed. Although some wall sections have been restored since the beginning of the communist era (1949), considerable amounts are still in poor condition. The Wall is often quarried for stones for constructions in the nearby villages; great amounts of stone are also used for road surfacing.

The Great Wall was built in various dimensions, by various techniques and from various building materials. However, a trapeze-shaped cross-section is apparent almost everywhere. To increase efficiency, along some sections outer trenches and ramparts were made. Around Beijing, the walls are 4–8 m wide and 7–8 m high, and watch-towers 12 m in height were erected at a few 100 m intervals. According to estimations, ca. 25,000 watch-towers were built along the Great Wall of China and communication was provided by further 15,000 signal-towers.

So-called earth mottes were built of loam (loess or clay); in more humid regions they were almost entirely erased by erosion. In the western sections, the wall was constructed by overlapping reed and tamarisk branches, as well as sand and gravel, followed by the strong compaction of these layers. Along some sections, loam, in order to improve its stability, was mixed with straw and rice. Facing sporadic rains and deflation, the walls are still 3 m high and 4 m wide at some locations after 2,000 years.

14.3 Defence in the Roman Empire: Limes and Earth Ramparts

In addition Late Bronze Age and Iron Age *earth mottes*, *ramparts* and the *rectangular Celtic earthworks*, defence constructions of the Roman Empire are also significant geomorphological features. Even the Roman legions participating in conquests established march-camps and temporary command posts fenced round with palisades. To secure the stabilised empire and hinterland necessary for further conquests, *fortified camps* (castrae, castellae) and watch-towers were established that were part of the limes, i.e. the centrally planned defence line of the Roman Empire. These military camps, mostly rectangular in shape and facing the enemy with their shorter side, were supported by trenches and palisades.

From a geomorphologic point of view, ditches and mounds along the *limes* are the most significant. One of the most spectacular in Europe, also part of the UNESCO World Heritage is the limes in Germany, morphologically well visible at many locations. The best preserved sections of this limes, 584 km in length, with about 1,000 watch-towers and 100 castellae protecting Upper Germany and Raetia from German attacks, are found in the Taunus Hills and along the northern border of Wetterau (Fig. 14.1). Here, the limes functioned as the border between settlements during the Middle Ages, thus they were further deepened and maintained for the protection of the given territories.

Hadrian's Wall, erected between 122 and 128 A.D. against the Picts of Scotland, is 120 km in length, and located near the border between Scotland and England. The *Antonine Wall* to the north between the Firth of Forth and the Firth of Clyde was built of stones and peat in 142 A.D.

Csörsz' Trench (at some locations, referred to as Devil's Trench) in the Great Hungarian Plain was built by the Sarmatians initiated by Emperor Constantine with help by the Roman Empire between 322 and 332 A.D. The trenches are 2–4 m deep and 5–9 m wide, were made by the Sarmatians. Embankments were erected with 7–10 m basal width and 2.5–3 m height. They include, at many locations, up to 4 quasi-parallel rampart lines running in a distance of 3–15 km from each other, surrounding the Great Hungarian Plain from the north and the east in a total length of about 550 km. It protects an area of approximately 60,000 km^2 protected (Fig. 14.2). To the development of the Csörsz' Trench, ca. 15 million m^3 earth had to

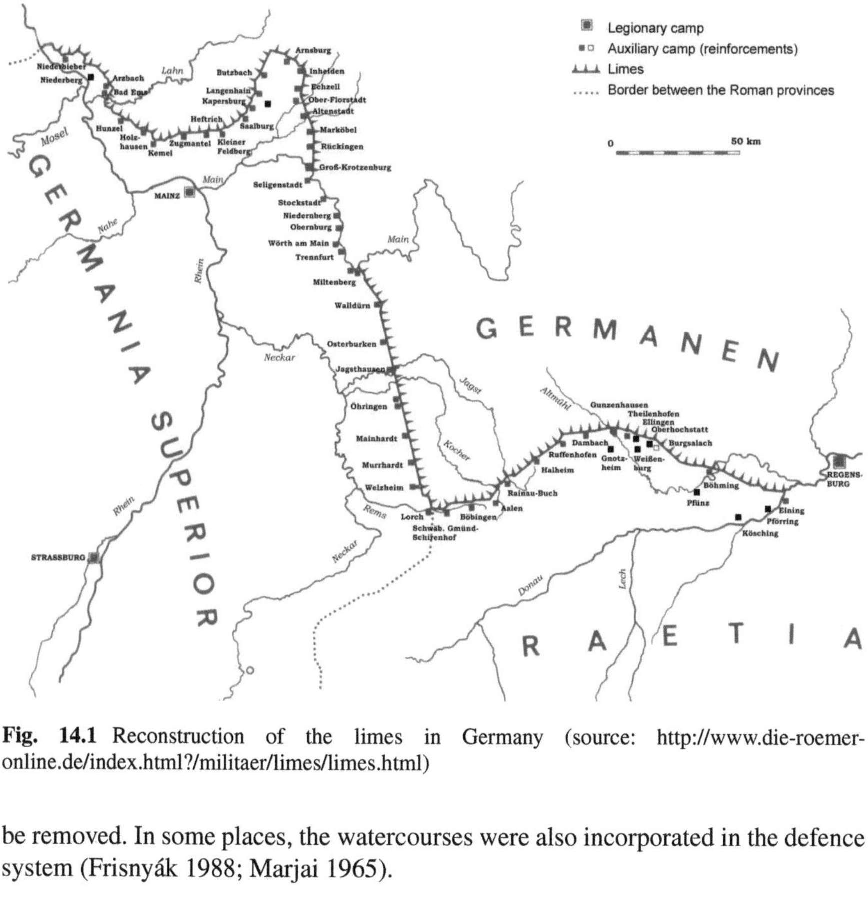

Fig. 14.1 Reconstruction of the limes in Germany (source: http://www.die-roemer-online.de/index.html?/militaer/limes/limes.html)

be removed. In some places, the watercourses were also incorporated in the defence system (Frisnyák 1988; Marjai 1965).

14.4 Medieval Ramparts and Fortresses

Medieval times witnessed the continuation of the construction of earth ramparts relevant for anthropogenic geomorphology. A remarkable monument is the earthworks of *Offa's Dyke*, 20 m wide and 2.5 m high, constructed along the boundary of the Kingdom of Mercia in England and Wales in the 8th century. Of the many Medieval ramparts and earth fortifications found in the Carpathian Basin, the *earth motte of Szabolcs* (Plate 14.1), a former administrative centre, is outstanding regarding both its dimensions and significance. To construct the ramparts of this earth motte, irregularly triangle-shaped with a length of nearly 800 m and a height of 15–20 m, but even of 25–30 m at some sections, a vast amount of timber was used. Its present volume is estimated to be 326,000 m^3 (Frisnyák 1990).

Rock castles, fortresses, citadels, watch-towers, stone walls and their ruins from the Middle Ages and the Modern Age are traditionally studied by experts of the protection of historical monuments and archaeology, among the elements of the built

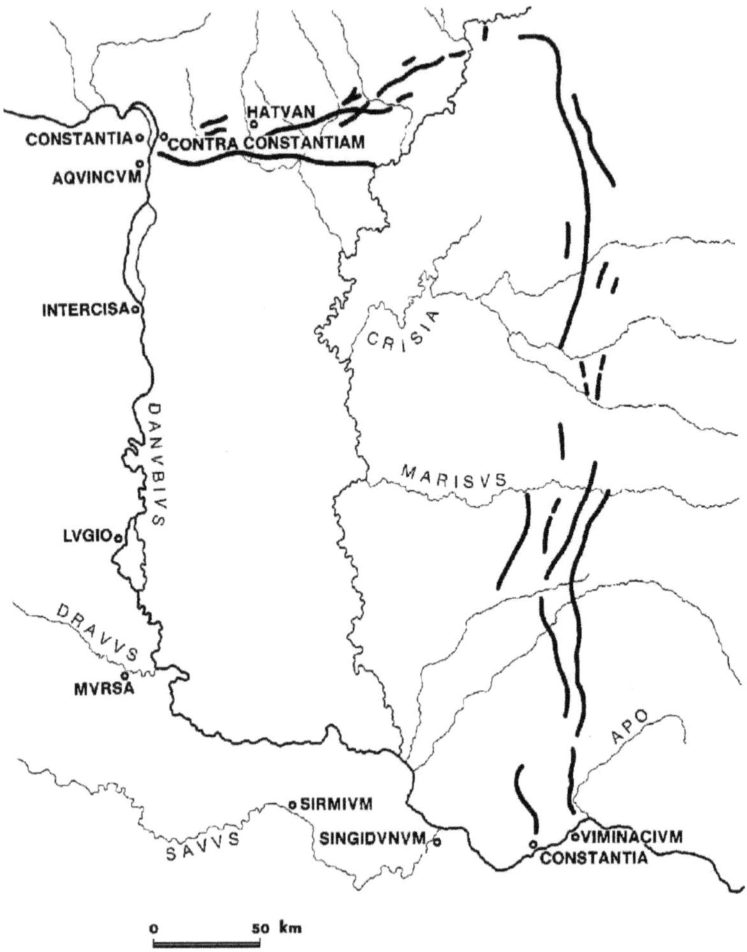

Fig. 14.2 Rampart systems in the Great Hungarian Plain in the 3rd and 4th centuries A.D. (Visy 1989)

cultural landscape. The study of mounds with the ruins of such structures, accompanying excavation landforms rampart systems and earth mottes are also within the sphere of study of anthropogenic geomorphology.

14.5 Earthworks in the Modern Age

From the point of view of anthropogenic geomorphology, the various *earthworks* of *the Modern Age* are also challenging. The primary military function of earthworks in the Modern Age was to shield against bullets. According to the directions of the enemy's fire-power, various designs were in use for earth entrenchments (e.g. epaulement, traverse, rear). The ramparts were made up by the gun post,

Plate 14.1 The earth motte of Szabolcs (source: http://jam.nyirbone.hu/)

the inner trench, the outer trench and the glacis (Fig. 14.3). Apparently, not all of these were entirely constructed as the structural complexity of the ramparts was influenced by the military situation and the time available. The identification of the original forms is made more complicated by ensuing geomorphic processes.

Redouts are 4–5-sided embanked ramparts giving protection against the fire of more powerful cannons. Their size depends on the power of besiegers and the number of cannons; they were usually constructed for 1–2 companies of infantry and 4–8 field cannons (Plate 14.2).

The most impressive example of the fortress architecture in the Low Countries is the Fortress of *Bourtange* (Plate 14.3) near Groningen. This fortification system constructed of earth exclusively was built during the Eighty Years' War, between 1580 and 1593 to the order of the Prince of Orange to isolate Groningen occupied by the Spanish. In front of the broad moat, a low dike (faussebraie) was erected; it was supported by a number of external ramparts (crowns), three-sided corner bastions /horn as well.

The most famous and productive planner of fortifications was *Marshall* Sébastien le Prestre de *Vauban* (1633–1707) of France, using the innovations of the previous centuries in defence systems being the most complex and diverse, geometrically precisely composed and the most adjusted to the geographical conditions. The best known among his 300 fortresses in France, is the *Neuf Brisach* (Neubreisach) in Alsace built between 1699 and 1703.

Among the earthworks of historical Hungary of great strategic importance, the *fortification system of Komárom* (Komarno) is worth of special attention. Of the Medieval and Early Modern Age fortresses reconstructed during the 19th century,

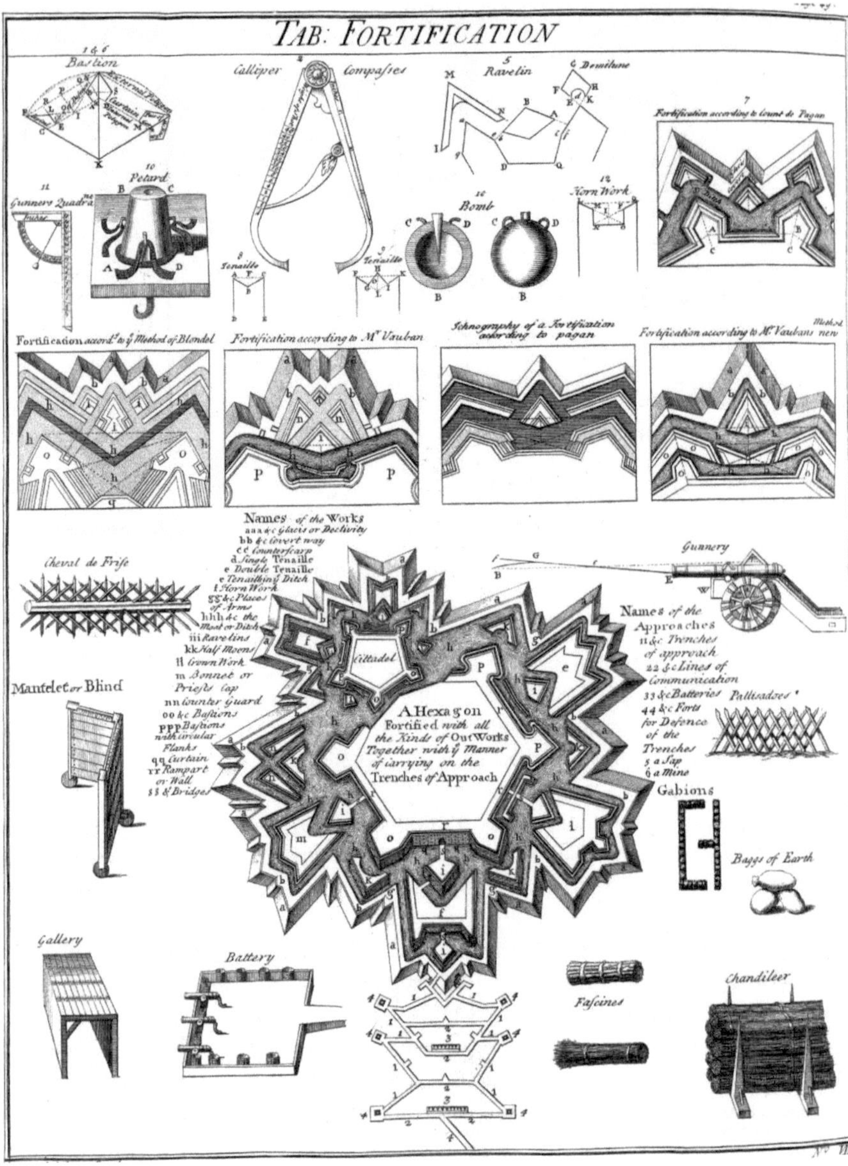

Fig. 14.3 Ramparts of the modern age (source: http://theudericus.free.fr/Vauban/Vauban.htm)

closed traverse-shaped Fort Monostor (Plate 14.4), built between 1850 and 1871, is considered to be the largest of modern Central European fortresses. This building complex covered by earth covers an area of approximately 70 ha and was constructed as a labyrinth of ramparts, bastions and underground casemates built of stone and brick (Gráfel no date).

Plate 14.2 Austrian redout from the 18th century in the Ghimes Pass, at the former Transylvanian-Moldavian boundary – today the County of Bacău, Romania (Ilyés 2004)

Plate 14.3 The Fortress of Bourtange (Sources: http://www.pixdaus.com/pics/1211860758iFwLnst.jpg)

Plate 14.4 Fort Monostor in Komarno (source: http://hu.wikipedia.org/wiki/F%C3%A1jl:Kom%C3%A1rom_Fortress_03.jpg)

Fortresses began to lose their strategic importance as early as the late 18th century. Fast, flexible invasion troops during the Napoleonic wars simply avoided the old defence systems. Fortifications were reconstructed primarily in large towns and strategic junctions of supply routes. Moreover, the major battlefields (Leipzig, Jena, Waterloo) also witnessed significant anthropogenic transformation.

14.6 Anthropogenic Geomorphological Impacts of World War I

The increasing fire-power, military technique innovations and positional warfare at many fronts lasting for years caused extraordinary military landscape transformations during World War I. The most spectacular example is the *battlefield of Verdun*. As a result of the massive artillery attacks, this battlefield was transformed into a cratered landscape within a few months. German troops alone fired 1,350,000 tonnes of grenades during the 30 war months and killed 1,000,000 people. On each hectare ca. 50 tonnes of splinters are buried underground even today. Following the seizure of *Fort Douaumont* (Plate 14.5) of key strategic importance by the Germans, the most intensive fights took place on the left bank of the Meuse River: during the battles around 'Dead Man's Dump' and 'Dump No. 304', ca. 40 million bullets and mortal shells were dropped (two on each square metre) to reshape the surface. 'Dump No. 304' deepened 16 m during the battles of spring 1916.

During the wars of the 20th century, especially due to the positional warfare in World War I, *trenches* and the various automatic rifle and machine gun posts established at battlefields and defence lines are remarkable monuments of human landscape transformation. Trenches provide shelter against machine-gun bullets

Plate 14.5 Pre- and post-battle aerial photos of Fort Douaumont (Sources: http://upload.wikimedia.org/wikipedia/commons/0/0e/Fort_Douaumont_Anfang_1916.jpg and http://www.freeinfosociety.com/media/images/3645.jpg)

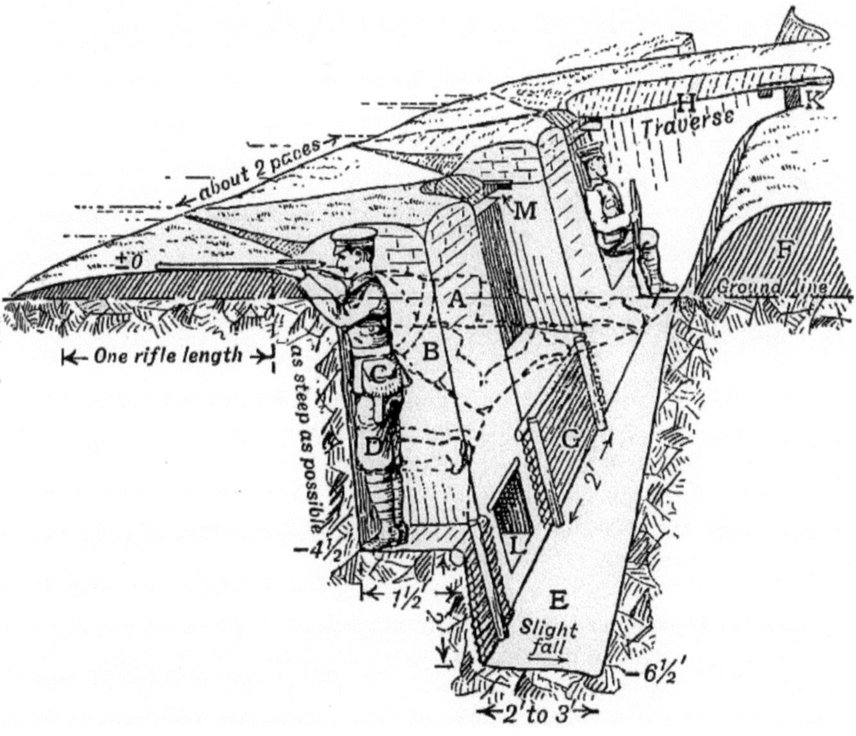

Fig. 14.4 The structure of a fire trench (1914) (source: http://upload.wikimedia.org/wikipedia/commons/e/ef/Trench_construction_diagram_1914.png)

and grenades and the opportunity for soldiers to securely fire, mostly in standing or kneeling positions (Fig. 14.4). The breastworks were originally supported by sandbags. Trenches often had a zig-zag planform in order to ensure better defence against grenades and enemy attacks. Such trenches are classified, according to their functions into operational, communication, cross, attack and reserve types; these functions has had an influence on the formation and texture of the trench systems still seen today.

The intense fighting between the Austro-Hungarian and Italian armies along the Italian front during World War I resulted in significant changes in the landscapes of the Southern Alps region (the Isonzo region, the Dolomites). Mountain roads were undermined. One of the best known symbols of battles is the 2,462 m high *Col di Lana* (Buchenstein) in the Dolomites that was blown up by Italian divisions on the night of 17–18th April 1916. About 5,000–5,500 kg explosives were stockpiled in two bomb bays; due to the explosion a dual crater, 25 m wide, 35 m long and 12 m deep was formed. Between 1916 and 1918, 34 similar explosions occurred along the mountain fronts of Tirol alone.

Bomb bays also formed on the western front. The limestone region along the *Somme River* was especially suitable for underground tunnel driving. The tunnel divisions of the British Army, mostly of civilian miners, drilled tunnels to place explosives. The front along the Somme witnessed two vast explosions beneath the German posts on 1st July 1916: one of them at La Boisselle and the other at Beaumont-Hamel (Hawthorn Ridge). At La Boisselle a 'crater' 30 m in depth and 90 m in width, still preserved today was formed.

Along with the development of military techniques, *road and railway construction for military purposes* also resulted in new landforms.

14.7 Landscape Transformation by Warfare in World War II and the Preceding years

Following World War I, defence lines, concrete-cored fortification and bunker systems (e.g. Maginot Line, Westwall or Siegfried Line) were constructed to counteract large-scale attacks and invasions across Europe. Further examples Atlantikwall and Ostwall, the fortification system constructed on the border of the Republic of Czechoslovakia by the Germans in the 1930s and the Árpád Line built in the Carpathians in the 1940s.

In addition to defence lines, more spectacular concrete and stone bunkers and fortifications of the modern age, traces of fire trenches (by today, partly filled up) as well as depressions of machine-gun nests, anti-aircraft batteries and tank-traps are still visible today. Gunpits were linked by trench systems. Remnants of the sites of field kitchens and signal posts surrounded by earthworks or planated as well as those of military barracks destroyed during military actions and in post-war times are similarly identifiable.

In the battlefields of World War II, the traces of impacts of bullets have been preserved. On maintained, regularly mown mountain meadows, microtopographic textures of depressions at the sites of grenade impacts as well as the flabellate-shaped ones of hummocks of explosion in origin at their opposite side are easily observable (Ilyés 2004; Rathjens 1979).

Interventions destroying *flood prevention facilities* and *dams of hydro-power plants* are considered to be among the geomorphological impacts of wars, or in a more direct sense, strategic planning and defence. During World War II, the Germans, in fear of the allied invasion, caused inundations by ruining dams and embankments along the coasts of Belgium and the Netherlands. In autumn 1944, the allies, by bombing the Westkapelle Dam, thought to be the most solid structure in Europe at that time, flooded the Isle of Walcheren. Such floodings for military purposes between 1940 and 1945 impacted about 230,000 ha in the Netherlands (Nir 1983; Dávid 1998).

Barrage dams on the rivers Möhne, Eder and Sorpe in North Rhine-Westphalia were attacked by special bombs by the Royal Air Force at the dawn of 17th May 1943. This intended to disable weapon manufacturing capacities in the Ruhr Region. With the exception of the one on the Sorpe River, all dams were damaged; the water sometimes 12 m deep caused serious damage in the neighbouring towns and killed about 2,000 people. On the 48 m high dam of the River Eder, a 70 m wide and 22 m deep hollow was formed following a bomb explosion, causing the flow 8,000 m^3 water per second (in total 160 million m^3). Due to the immense amount of water running down, hundreds of houses, factories, railway lines, roads and bridges were damaged or drifted away, and a vast lake was formed at the estuaries of the Schwalm-Eder and the Eder-Fulda, and in the Kassel Basin (http://de.wikipedia.org/wiki/Edersee).

Rubble heaps (Trümmerberge, Schuttberge) raised following the bombings in World War II are found in many European cities (Szabó 1993). Only in Germany, in 1945, 400 million m^3 of rubble had to be removed. Such rubble mounds in Berlin include the 15% of the total amount of rubble in Germany. In the Park of Friedrichshain, two rubble heaps piled on bunkers are found, i.e. the kleiner (smaller) and grosser (greater) Bunkerberg. In these two 'hills', approximately 1 million m^3 of rubble was piled.

14.8 Geomorphological Impacts of Modern Wars

Modern wars, due to the transportation of considerable amounts of ammunition, transform the surface over large areas. Mainly military roads and air bases are involved as well as degradation caused by military vehicles should be mentioned. Constructions carried out by the American army in Vietnam, made soil exposed at large areas resulting in significant rates of erosion. On the other hand, the slopes and stepped slopes resulting from construction works were further shaped by monsoon rains and deflation (Nir 1983).

Between 1963 and 1971, in South Vietnam and in the neighbouring Laos and Cambodia, the American forces sprinkled herbicides estimated to be 72 million litres (of which 44 million litres was the chemical Agent Orange containing dioxine, still causing severe health problems). On 17.8% of South Vietnam's area (i.e. 3,004,000 ha), the primary aim of actions to burn the foliage of forests and mangrove forests was to eliminate shelters, supply bases and food-producing agricultural areas of partisans. The large-scale deforestation resulted in inundations in the rainy season and soil desiccation and deflation in the dry season. Soil eroded from areas of high relief causing the formation of laterite surfaces where forests could hardly rejuvenate. In lowlands, accelerated deposition of fine alluvial sediments and silt further increased flood hazard (Le Cao no date).

As a consequence of the technical progress in warfare, the increasing role of flexible, mechanised troops and the spreading of air warfare, 'position warfare' based on fire trench and bunker systems known from World War I has became outdated in the mid-20th century. Another reason for this was that during World War II, such fortification and bunker systems fell short of expectations and were easily passed by invasion forces. Notwithstanding, some examples of military actions based on trenches and bunkers can be found from the second half of the 20th century. The war between Iraq and Iran in 1980–1988, as a whole, was similar to the trench battles experienced during World War I, as both forces had significant artillery but a relatively small number of armoured divisions. Defence by trench lines did not prove to be sufficient during the First Gulf War following the occupation of Kuwait. With support provided by the air forces and armoured bulldozers, such defence lines (http://de.wikipedia.org/wiki/Grabenkrieg) were simply buried by the American forces. Similar trench war took place between Ethiopia and Eritrea. Today, ceasefire lines secured by traditional trench systems and bunkers are found between South and North Korea, and between Pakistan and India (in Kashmir).

The most accelerated and greatest geomorphologic transformations of modern warfare were caused by the testing of *nuclear weapons*. In addition to the bombing of Hiroshima and Nagasaki, which caused damage unimaginable so far, between 1945 and 1998, 2057 nuclear weapon tests were carried out. The fire power of all nuclear power testing carried out to date equals that of 1 billion tonnes of TNT. Of the 2,057 nuclear weapon explosions, the total fire-power of the 1,529 underground explosions was ca. 82.5 mega-tonnes. The first nuclear bomb was blown up on 16 July 1945 under the supervision of Robert Oppenheimer in Alamogordo (New Mexico); following the explosion of this 19 kilo-tonne bomb, a crater 800 m wide and 3 m deep was formed while the desert sand melted into a greenish-coloured glass. This crater was filled up in the 1960s. One of the largest crater resulted from the Sedan test carried out on 6th July 1962 at the test site in Nevada. The 104 kilotonne nuclear bomb was emplaced in a depth of 193 m. This explosion removed 12 million tonnes of earth and left behind a crater 390 m in diameter and 97 m in depth (Goin 1991).

References

Dávid L (1998) Antropogén geomorfológiai tanpéldák a középiskolai környezetvédelmi nevelésben. (Anthropogenic geomorphologic study cases in the environmental education on the secondary school level). In: A környezeti nevelés és a földrajztanítás. (Segédanyag a helyi tantervek készítéséhez a Cholnoky pályázatok alapján). Földrajztanárok Egylete – Cholnoky Jenő Alapítvány – Flaccus Kiadó, Budapest

Engels F (1859) Fortifikation. In: Marx K, Engels F (ed.)(1972) Werke. (Karl) Dietz Verlag, Berlin. Band 14.

Erdősi F (1969) Az antropogén geomorfológia mint új földrajzi tudományág (Anthropogenic geomorphology as a new discipline of geography). Földr Közl 93 (1): 11–26

Fehn K (1997) Konversion militärischer Liegenschaften als Aufgabenfeld der Kulturlandschaftspflege. In: Schenk W, Fehn K, Denecke D (eds.), Kulturlandschaftspflege. Beiträge der Geographie zur räumlichen Planung. Gebrüder Borntraeger, Berlin-Stuttgart

Frisnyák S (1990) Magyarország történeti földrajza (The Historical Geography of Hungary). Tankönyvkiadó, Budapest

Goin P (1991) Nuclear Landscapes: Baltimore. John Hopkins University Press, Baltimore

Gráfel L. (no date) Nec arte, nec marte. A komáromi erődítményrendszer. (Nec arte, nec marte. The Fortification System of Komarno). http://www.komarno.sk

Hard G (1973) Die Geographie. Eine wissenschaftstheoretische Einführung. Walter de Gruyter, Berlin-New York

Ilyés Z (2004) A hadászati-védelmi objektumok rendszere és történeti rétegzettsége a Gyimesi-szorosban (The system and historical exfoliation of warfare-defence objects in the Ghimes pass). In: Frisnyák S, Csihák Gy (eds.), Gyepűk, várak, erődítmények és egyéb honvédelmi létesítmények a Kárpát-medencében (895–1920). Nyíregyházi Főiskola Földrajz Tanszék Nyíregyháza-Zürich

Le Cao Dai (no date) Die Auswirkungen von Agent Orange und der Fonds für die Opfer von AO. http://www.beepworld.de/members34/der_rest/agent_orange.htm

Marjai M (1965) Ördögárok a debreceni Nagyerdőn. (Ördög trench in the great forest of Debrecen). A debreceni Déri Múzeum évkönyve 1962–1964. A Debreceni Déri Múzeum Kiadványai 47: 87–92

Mortensen H (1954/1955) Die quasinatürliche Oberflächenformung als Forschungsproblem. Wissenschaftliche Zeitschrift der Ernst Moritz Arndt-Universität Greifswald, Mathematisch-naturwissenschaftliche Reihe IV. 6/7: 625–628

Nir D (1983) Man, a Geomorphological Agent. Keter Publishng House, Jerusalem

Rathjens C (1979) Die Formung der Erdoberfläche unter dem Einfluß des Menschen. Grundzüge der Anthropogenetischen Geomorphologie. B. G. Teubner, Stuttgart

Schlüter O (1926) Die Urlandschaft. In: Volz W (ed.), Der ostdeutsche Volksboden. Aufsätze zu den Fragen des Ostens. Ferdinand Hirt, Leipzig

Szabó J (1993) A társadalom hatása a földfelszínre (antropogén geomorfológia). (The impact of society on the Earth's surface – Anthropogenic geomorphology). In: Borsy Z (ed.), Általános természeti földrajz (Physical Geography). Nemzeti Tankönyvkiadó, Budapest

Visy Zs (1989) A római limes Magyarországon (The Roman limes in Hungary). Corvina, Budapest

Website Sources

http://www.die-roemer-online.de/index.html?/militaer/limes/limes.html

http://de.wikipedia.org/wiki/Edersee

http://www.pixdaus.com/pics/1211860758iFwLnst.jpg

http://de.wikipedia.org/wiki/Grabenkrieg

http://jam.nyirbone.hu/

http://hu.wikipedia.org/wiki/F%C3%A1jl:Kom%C3%A1rom_Fortress_03.jpg

http://upload.wikimedia.org/wikipedia/commons/0/0e/Fort_Douaumont_Anfang_1916.jpg and http://www.freeinfosociety.com/media/images/3645.jpg

http://upload.wikimedia.org/wikipedia/commons/e/ef/Trench_construction_diagram_1914.png

http://theudericus.free.fr/Vauban/Vauban.htm

Chapter 15
The Impact of Tourism and Sports Activities

Lóránt Dávid, Zsuzsanna Lontai-Szilágyi, and Zoltán Baros

Abstract Though recreation activities exert a rather limited influence on the environment in comparison to other sectors like industry and agriculture, in planning more attention should be paid to their environmental impacts (particularly on topographic conditions). This impact on the relief can be manifold. The establishment of ski-trails starts with deforestation and continues by bulldozing or blasting for levelling the landscape. Waterside environments are also popular scenes for tourism and sports. Finally, the impacts of golfing, motor sports, hiking (camping, rock collecting, etc.) are also mentioned with examples.

Keywords Recreation · Skiing · Water sports · Tourism development · Off-road vehicles · Erosion

15.1 Introduction

As sport-oriented travel is gaining more and more ground, research into the impact of tourism and sports on the physical environment is increasingly significant. For example, in Germany 55% of all travel (32 million citizens) is motivated by sport whereas this is 52% in the Netherlands (7 million citizens). Sports are a less significant motivation for travel in France, although even there 23% of all travel (3.5 million citizens) is related to some kind of sports activity (Ritchie and Adair 2002). This remarkable trend is not restricted to Europe.

Apart from travel based on natural attractions and sports, the range of destinations and that of outdoor activities as well as the length of the time devoted to them are also increasing (Strasdas 1994). Moreover, the number of travel organizers offering free-time sports programs to their travellers also indicates an

L. Dávid (✉)
Department of Tourism and Regional Development, Károly Róbert College, 3200 Gyöngyös, Hungary
e-mail: davidlo@karolyrobert.hu

J. Szabó et al. (eds.), *Anthropogenic Geomorphology*,
DOI 10.1007/978-90-481-3058-0_15,

increase (Standeven and Knop 1999). According to surveys by the World Tourism Organization, the most popular physical activities among recreational travels are skiing, snowboarding, climbing, touring, water sports and cycling (WTO 2001). The popularity of outdoor activities in certain destinations may lead to high concentrations of the participants in both time and space, resulting in significant infrastructure development, involving even higher pressure on the environment.

The impacts of recreational tourism and sports are traced in the physical environment of the areas affected and depend on several factors (Fig. 15.1).

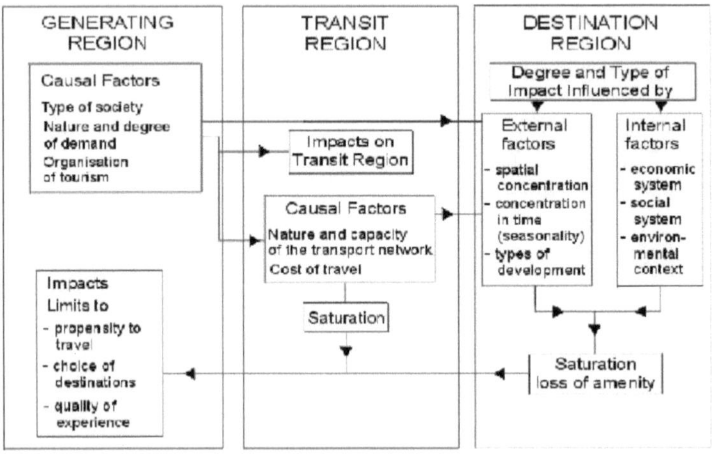

Fig. 15.1 The relationship between the causal factors and impacts with the generating, transit and destination areas (after Pearce 1989 after Thurot 1980)

Consequently, although the *types of impacts* are *similar* in all affected areas, there are differences in their *exact way of appearance, extent and bulk* (Puczkó and Rátz 2002). Meanwhile, we have to remember that the sites for recreational tourism and sports are usually *rather complex and susceptible environments* (e.g. small islands, coastal areas, mountainous areas).

The exploration of the entire physical impact system of recreational tourism and sports activities is a rather difficult task. Factors rendering the impact assessment more difficult are as follows:

- The consequences are often present in conjunction with those of other (industrial, agricultural, transportation, etc.) activities and they are hardly separable.
- Researchers do not have sufficient knowledge on the conditions of the environment of the target area prior to the appearance of tourism and/or sports activities; therefore, they lack the basis for comparison.
- The direct impact is often coupled with indirect or long-term transformations.
- As a result of the complicated interrelations, consequences are not always observed at the same sites where the impact itself is manifested.

– During comparative evaluations or preparing comprehensive system-approach studies, the various impacts originating from the difference in the supply and demand factors represent a great difficulty (Marton-Erdős 2001).

Processes, links related to the activities mentioned above, causing the transformation of landforms of the Earth's surface, creating new landforms as well as altering or removing existing ones, are presented below.

15.2 The Impact of Leisure-Time Travels and Sports on Relief

The mostly deleterious impacts of tourism and sport on the physical environment have been studied by several researchers. Based on their findings, it is claimed that such activities play a major part in *landscape (topographic) transformation related to the establishment of infrastructure* and in *accelerated erosion.*

Most outdoor tourism and sports activities require major infrastructural investments (transportation facilities, quarters, catering, sports fields, various service facilities). Thus, *the establishment of infrastructure results in significant transformations in the environment (topography) of the target area.* The pressure involved in the intervention is aggravated by the fact that often *pristine natural areas* are affected (turned into recreation complexes, golf courses, ski runs).

As seen on Fig. 15.2, there are only a couple of outdoor activities, which do not call for an infrastructural background with profound intervention into the landscape.

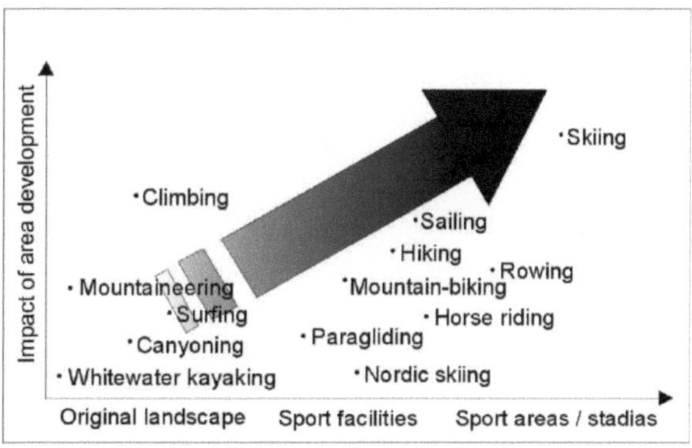

Fig. 15.2 The relationship between the area, the features of activity and the intensity of impacts (after Türk et al. 2004)

15.2.1 Impacts from Winter Sports

The favourite destinations of winter sports with worldwide popularity are the Alps, the Pyrenees, the Scandinavian Mountains, the Carpathians, the Andes, the Rocky Mountains, the Himalayas, the Caucasus and the Southern Island of New Zealand.

Initially, skiing was a mass phenomenon restricted to the Alps, where the first undesired impacts were observed. The snow-covered areas of these ranges are visited annually by about 4050 guests enchanted by downhill skiing. They are served by ca. 15,000 ski-lifts and 40,000 ski-trails (with a total length of ca. 120,000 km) (Marton-Erdős 2001). During the establishment of ski-trails, the affected area is first deforested and *steep slopes* are *planated* by bulldozing or blasting whereas *gentle slopes elevated* or elsewhere the *landscape levelled*. The deforestation results in reduced water retaining capacity and in *accelerated soil erosion*. Rills or, in extreme cases, gullies and ravines are cut into the surface of bare slopes by rapid meltwater and rainwater runoff, although sheetwash is common even on levelled and compacted surfaces. With the winter passing, unhindered runoff on treeless slopes may commonly cause *flash floods coupled with landslides* threatening facilities and even human lives. For instance, in Tirol in July 1987, a disastrous landslide left 60 people dead and 7,000 citizens of 50 settlements homeless (Romeril 1989). Mass movements due to the establishment of tourist infrastructure have been described from Italy, Switzerland and Nepal (Holden 2000). For mass tourism, the Alps have become the most attractive site in the world. Due to the construction of hotels and infrastructure to ski runs and restaurants, the landscape has undergone dramatic changes and has become fragmented. According to Mosimann (1985), the share of tourism-induced transformed areas in the Swiss Alps may be up to 15%. During his research on ski-trails in Switzerland, he found that the rate of erosion is mainly determined by the shape of slope, soil moisture content, runoff frequency and catchment area.

Alterations due to winter sports may be present elsewhere too. Urban environments are often transformed in order to fulfil the demands of visitors. For example, the ski-jump of Oslo is apparently a dominant feature of the townscape. Figure 15.3 indicates how the original slope has become steeper during a decade.

Urban ski-trails and the surrounding complexes attract several thousands of tourists each year (Bale 1989). As an example from Hungary, mounds for sledging in numerous lowland settlements (in Debrecen, among others) could be mentioned.

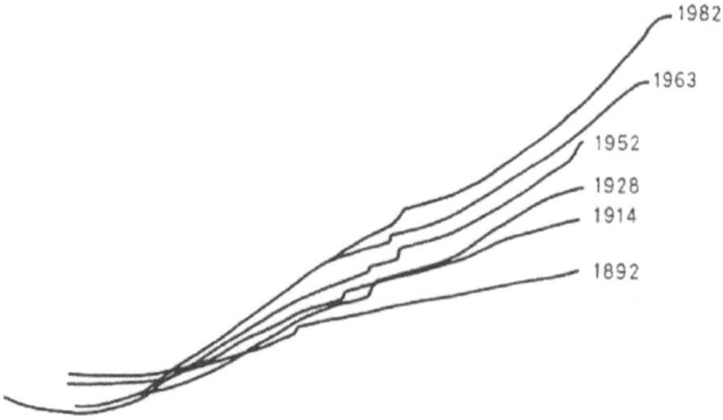

Fig. 15.3 Changes in the inclination of the ski-jump 'Holmenkollen' between 1892 and 1982 (after Bergsland and Seim-Haugen 1983)

To cite a positive experience, on the slopes of the Cairngorm Mountains in Scotland where new chair ski-lifts have been put into operation, the rate of soil erosion has significantly dropped due to turfing and forestation in the areas of former outdated ski-lifts as well as establishing drainage ditches (Bayfield 1974).

Grass-skiing is not a winter sport but originated in skiing on snow. The fans of skiing, nowadays, are not satisfied with the short winter skiing seasons, therefore this type of skiing without snow has become more and more popular. Tivers (1997) claims that trails are often established in England in reclaimed coal mines or other industrial sites. Nine grass ski-trails are found just in Sheffield, their total length being ca. 1,000 km. As a consequence of the direct and intensive pressure on vegetation and slopes, grass-skiing has even more severe impacts than traditional skiing as far as landscape destruction is concerned.

15.2.2 Impacts Near Water

Apart from mountainous areas, natural water surfaces and waterside environments are also popular scenes for tourism and sports. At the time of spreading mass tourism, the most important destinations for summer holidays possess the attractions of the 'Three S's (Sea, Sand, Sun). They are coasts and subordinately lake shores and river banks.

Most *coastal areas suitable for bathing* have been the victims of *the development of various service facilities*. During the construction works, *sand dunes* and protective coastal vegetation are removed and intensified *coastal erosion* reduces coastal stability. Environmental experts claim that the mangrove forests of Southeast Asia that have been deforested to provide ground for coastal paradises for tourists, among others, might have helped to withstand the tsunami of 26th December 2004, which demanded more than 200,000 victims and caused huge damage. Since the disaster, mangrove forests are declared to be protected in many Asian countries, as well as the natural coastal protection is attempted to be revived by re-planting. Similar phenomena were described in connection to the destruction of coral reefs (Figs. 15.4 and 15.5) at the coastal areas of Kenya (Holden 2000), providing a complex outline not only of the problems caused by erosion but also those due to tourism development.

Since 1960, almost three-fourths of the sand dunes in the near-shore areas between Spain and Sicily have disappeared (ECOSOC 1999). As seen from Table 15.1, coastal erosion affects a significant percentage of the Mediterranean coasts. Destruction of the foreshore is a major problem for 18% (in average) of the coastal areas with a total length of 18,000 km. Coasts of islands are especially susceptible to devastating forces. The rate of wave erosion is indicated by an estimate according to which Greece loses one island each year.

In course of time, as a result of coastal erosion, tourist facilities themselves require protection, thus *protective walls* are built, causing further damage to the landscape. Protection facilities are very common, e.g. on the coasts of the North Sea. Another, frequently applied way of protection is the *infilling of coasts* (Fig. 15.6).

Fig. 15.4 The coast before tourism development (after Holden 2000)

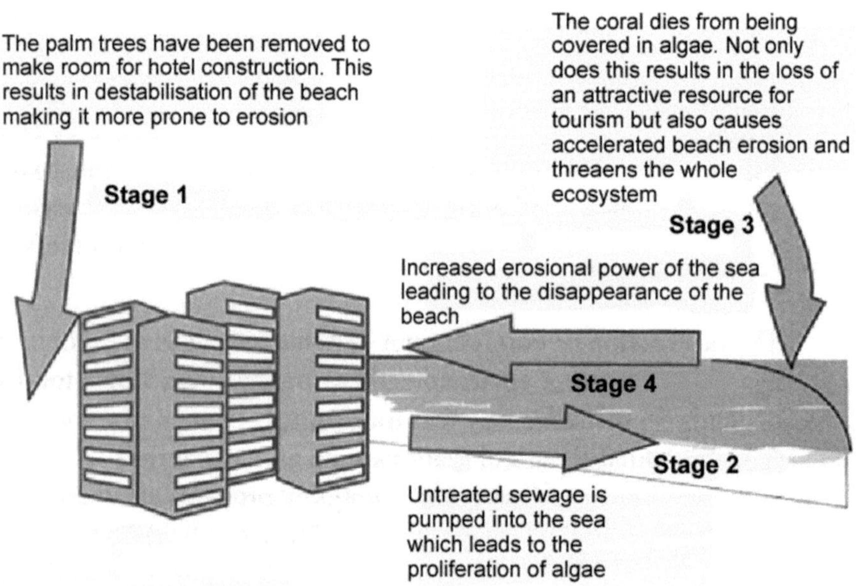

Fig. 15.5 The impact on coasts after tourist development (after Holden 2000)

Several investments have been made recently to increase incomes from tourism through *artificial archipelagos* constructed into the sea. The intricate system of the

Table 15.1 Development tendencies of the Mediterranean shorelines (CORINE Coastal Erosion 1998)

Coastal area	Total shoreline (km)	Stabilization (%)	Erosion (%)	Sedimentation (%)	No information available (%)	Unsuitable for recreation (%)
Balearic Islands	2,861	68.8	19.6	2.4	0.5	8.7
Gulf of Lion	1,366	46.0	14.4	7.8	4.1	27.8
Sardinia	5,521	57.0	18.4	3.6	16.0	5.0
Adriatic Sea	970	51.7	25.6	7.6	3.9	11.1
Ionian Sea	3,890	52.3	22.5	1.2	19.7	4.3
Aegean Sea	3,408	49.5	7.4	2.9	37.5	2.6

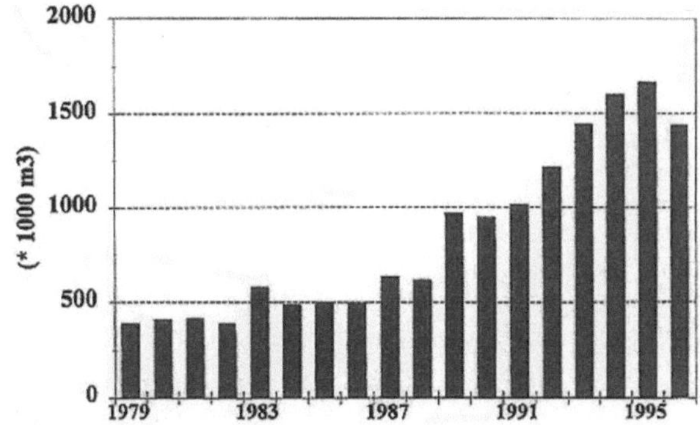

Fig. 15.6 The amount of sand and gravel used for the infilling of coastal areas in Belgium (after Haelters and Jacques 1996)

offshore physical environment has been encroached upon to an extent never seen before. In the United Arab Emirates, an artificial archipelago called 'The Palm, Jumeirah' is constructed off-coast. On its trunk, 8 km in length, luxury hotels, shopping centres, cinemas and pleasure grounds are placed; as are dwelling houses and offices on its 17 palm-branches. However, the most exclusive holiday resort of the Emirates will be a project consisting of 250 artificial islands, located 5 km from the coasts of Dubai, and representing the whole globe. Services of the archipelago called 'The World' are available for guests from 2009 (see http://www.palmsales.ca; Plates 15.1 and 15.2).

Turning to Hungary, at Lake Balaton, bathing culture started in the second half of the 19th century that, as a result of the loss of the coastal and mountain holiday resorts, was growing since World War I. Fulfilling the demands, land owners allotted their lands; by the construction of villas still seen today, developing the shores of Lake Balaton began in the first half of the 20th century. By the mid-1970s, 80% of the shoreline was built up in a 300 m wide zone. The level of human interference is indicated by the fact that only 70 km of the shoreline has remained in seminatural conditions, as opposed to the 195 km just after the millennium (Marton-Erdős 2001). When purchasing some of the properties 20–30 years ago, a marsh area could be found in the area of the present-day plots, of which infilling took place under

Plate 15.1 The artificial archipelagoes of The Palm, Jumeirah, located in Dubai (source: http://www.eikongraphia.com/?p=1868)

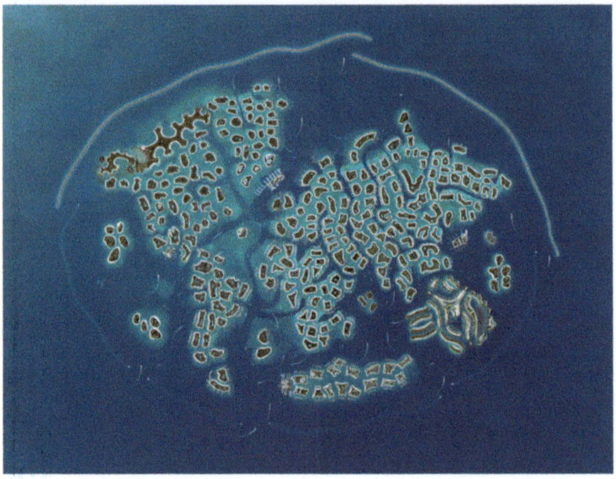
Plate 15.2 The artificial archipelagoes of The World, located in Dubai (source: http://www.eikongraphia.com/images/TheWorldMasterplanApr07CopyrightNakheelS.jpg)

strict supervision and according to plans approved by authorities. There were, however, many land owners intending to enlarge their plots along the shore illegally, by *infilling the Lake Balaton shore*. According to the regulations of the Lake Balaton Act of 2000 (62/2000 Act on the acceptance of Lake Balaton Primary Recreational Area territorial organization planning and the definition of the Balaton territorial organization regulations), protection is provided to the biota of the lake, on the one hand, by drawing a *new shoreline* (at some places by the infilling or levelling of shore areas), and to the stopping of the degradation of water quality threatening Lake Balaton as a tourist attraction, on the other.

15.2.3 Impacts of Golf Courses

Originatingin Scotland, golf has become a loved and popular sport. Due to its positive impacts on economy (e.g. above-average staying time, reduction of seasonality, high specific expenditures by tourists), it is becoming a more widespread product of tourism. Approximately 60 million people play golf annually on about 30,000 golf courses worldwide, with their total area larger than Belgium (Hodson 1996). In the early days of golf, natural topography was used to create smaller courses. The optimal environment was represented by hummocky landscapes, natural bunkers, soft lawn and sufficient drainage conditions. With technical development and market prosperity, the number and extent of courses kept on increasing, thus mountainous and coastal areas with extremely valuable ecosystems, deserts, derelict opencast mines as well as arable lands are being utilized (Bale 1989). When building up-to-date golf courses, the landscape undergoes considerable transformation, mainly through earth removal. For example, during the establishment of the Mária Valley golf course at Alcsútdoboz, Hungary, opened in 1997, 300,000 m^3 of soil was removed (Varga 1999). According to Wheat (1995) there are few ways of recreation more environmentally sound than golf. Although, he admits that growing concerns with this sport are gaining ground because of the large demand for land. The need to reduce environmental impacts from the allocation, establishment and operation of courses (e.g. creating artificial landscapes, increased fertilization and water use) was recognized in some European countries as early as the 1980s. Designs intended to reduce the size of the courses and to develop techniques to manage the physical environment in a sustainable way (Plate 15.3).

Plate 15.3 A golf course, located in Florida, enriched in artificial landscape elements (source: http://www.westchasegc.com/sites/images/442/Westchase-Golf-Club-FL-hole-13.jpg)

15.2.4 Impacts Related to Motorsports

Motorized sports have recently gained a worldwide and ever growing popularity resulting in the availability of a wide range of activities that can be classified into the following categories:

- Water or land (off-road or concrete track) motorsports
- Seasonal or all-the-year-round motorsports

For both primary (resulting from the activity itself and from the establishment of the necessary infrastructure) and secondary (induced by the activity), impacts on the surface should be distinguished.

The surface-forming impact caused by the constructions related to motorsports with tracks resembles to the consequences of road construction. Topographical disequilibrium occurs in a relatively small area, mostly manifested in reduced stability in the landscape, by the establishment of unusually steep and long slopes thus increasing erosion. The earthworks of roads – in addition to being surface- and landscape-forming factors en masse – also have an indirect impact on topography and microclimate (Erdősi 1987).

Recently, off-road driving (also referred to as *Toyotarization*) has been also seen as triggering sand and dust storms becoming more frequent in desert areas, as safari programmes with jeeps can also contribute to the disruption of the surface (Goudie 2004).

Additional damage is made by visitors' trampling, access roads, facilities and the construction of the associated infrastructure. Damage could be especially serious at times of races (concentrated use) where spectators leave the established and designated areas, e.g. to take photographs.

The degree of biophysical impact will depend on the location, intensity and duration of the activities themselves. Environmental response will vary in accordance to individual species/ecosystem resistance and resilience.

The amusement offered by off-road driving is now available to a greater number of participants. For them, the feeling of freedom, driving at illegal locations is of high priority regardless of its consequences being widespread as well as making the management of such impacts rather difficult.

15.2.5 Impacts of Green Areas

Landscape formation, *surface modification* of *diversified ways and extent* can be associated with the establishment of parks offering various ways of recreation. Just recall, for example, the public parks in cities with artificial water surfaces and mounds.

Another type of parks of human origin is *Olympic parks* for sport events. For example, in Munich, the rubble heap of Schuttberg, 60 m in height, was built for recreation purposes. In 1972, an area of 270 ha around it was developed and became the site for the 20th Summer Olympic Games, since then being one of the city's popular sport parks.

On the world's tourism market, *theme parks* offering all-day family entertainment are in fashion. Even the establishment of such amusement parks alone can have a considerable impact on the landscape by transforming the relief. In the meantime, due to land reclamation associated with archaeological parks, nature conservation areas, protected natural values, advantageous changes concerning the landforms can also take place.

When developing for recreational purposes, developers and planners are increasingly aware of the fact that in the long term, an indispensable base for recreation in nature is an attractive landscape preserved in its original shape. It is a widespread practice throughout the world that *investments aim at the restoration of the pristine physical environment.* A good example of habitat reconstruction with relief transformation is the area of the Hortobágy National Park in Hungary. In 2002, supported by a European Union's LIFE-Nature tender, it became possible to reconstruct the original state of a diversified, contiguous system of arid saline barren and marsh habitats. Within the framework of this *land reclamation programme* affecting nearly 8,000 ha, dams and channels restraining the natural surface water movements have been removed and flattened into the natural surface by *grading* (see http://www.hnp.hu).

As part of the landscape rehabilitation following the hydro-power project on the banks of the River Danube in the Visegrád area, in addition to the 20,000 m^3 land already reclaimed, a further 100,000 m^3 is planned to be filled for recreational purposes (see http://www.visegrad.hu).

Short-sighted economic interests may dominate over environmental protection even today, as in Hévíz where instead of landscape rehabilitation, a monster-sized hotel with 10 swimming pools is planned to be built. It was evincible earlier that the discharge of the lake source had been drained by the pumps of the hotels.

15.2.6 Recreation Activities Inducing or Increasing Erosion

Along pathways and in areas avoided by marked tourist routes, *treading* is a universal problem leading to the *soil compaction* and *erosion.* As a result of compaction, porosity and permeability are reduced and runoff increased. Water easily entrains soil particles and erodes soils. Meanwhile, as an effect of high pressure, the herbaceous plant cover protecting the soil becomes interrupted very soon and, during rainfall events, surface of the soil is sogged. By walking around the sections poorly passable, the widening of the paths starts, on the one hand, whereas soil erosion speeds up to a significant extent on unprotected surfaces (Fig. 15.7). This, in many cases, leads to creating a covered road surface.

With the help of the Wischmeier and Smith (1978) equation (Universal Soil Loss Equation), soil loss from areas affected by recreation activities can be calculated.

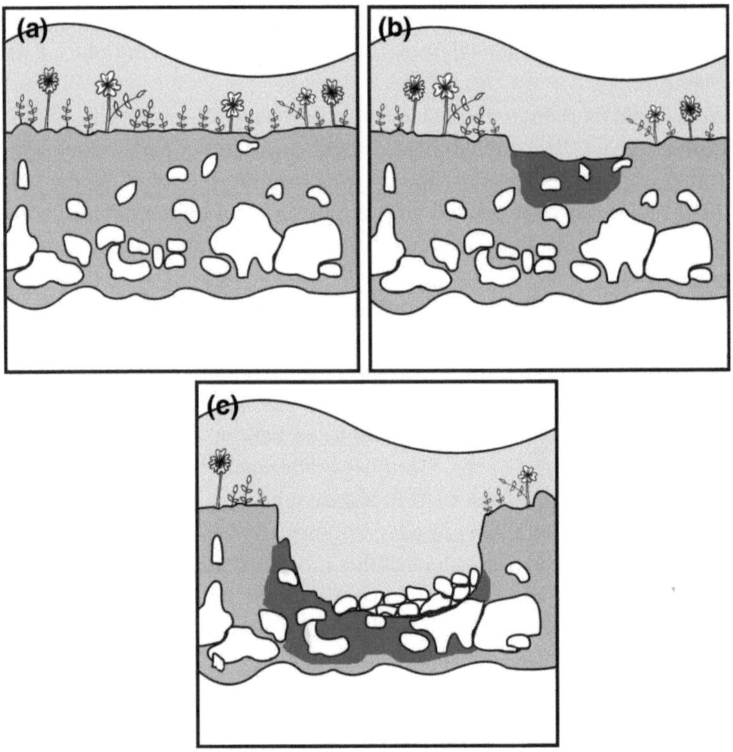

Fig. 15.7 The process of footpath erosion (re-edited after http://www.maridadi.com/related/feros.htm)

Links among the factors, however, have not yet been described properly. Leung and Marion (1996) extend the *factors influencing soil erosion* to some extent, and classify them according to the rate of their influence. They claim that the rate of erosion is *primarily* determined by *climate, geological, lithological settings and the type* and *intensity of the recreation activity*, whereas *topography, characteristics of the soil, vegetation cover* and *the attitude of users* play only a *minor* part in the process. The importance of reviewing the type of activities is illustrated in Table 15.2, indicating the degree of pressure on soils induced by the most characteristic ways of recreation from the aspect of treading.

Table 15.2 The impact of recreation activities on the soil

Activity	Pressure on soil (g/cm^2)	Source
Skiing	28	Liddle 1997
Touring (on foot)	297	Liddle 1997
Mountain biking	2,008	Eckert et al. 1979
Horse-riding	2,800	Lull 1959

Most experts (e.g. Dale and Weaver 1974; Deluca et al. 1998) agree that – disregarding motor vehicles – the most significant footpath erosion is caused by *horse-riding*, as the heels of horses put much more pressure on the soil surface (the aggregate weight of the horse and horseman can be up to 500 kg), than skis, boot-soles of hikers or the wheels of a bicycle. Phillips (2000), based on his investigation in the D'Entrecasteaux National Park in West Australia, claims that the rate of bare soil surface has grown from 5.2 to 31% following the crossing of 300 horsemen. In case the trace is the same, horse-riding rather deepens the path than widens it (Harris 1993).

The most remarkable impact of erosion caused by recreation activities can be observed in *highlands*. These areas, because of their steep slopes, shallow soil cover, especially at regions receiving more precipitation are extremely susceptible. Under a colder mountainous climate at a higher elevation above sea level, the vegetation is more vulnerable and tends to need a longer time for revival. Despite these unfavourable endowments, high mountains are under enormous demand. Following snowmelt, it becomes visible that long, bare soil strips run downslope – downhill tracks. The damage caused by grass-skiers who do not consider snow conditions, however, is not significant yet.

In mountainous areas footpath erosion caused by hikers is an increasingly relevant environmental problem. Karancsi (2000) claims that in the nature conservation area of Britain's Lake District attracting approximately 14 million visitors each year, footpath erosion often leaves behind huge scars in the landscape observable even on satellite images. According to the observations carried out in several national parks in Britain, when the angle of slope of a tourist route reaches 10 degrees following the destruction of the natural vegetation cover, soil erosion starts on the surface. Other highland areas suitable for hiking are obviously exposed to similar serious impacts (Plate 15.4).

The impacts of the use *off-road vehicles* (motorcycles, four-wheel-drive vehicles, etc.) on the surface should be investigated in detail (Fig. 15.8). The use of off-road vehicles and motorcycles, by creating or widening of trails (non-system routes), has an impact on soils, vegetation, wildlife and social conditions.

The ensuing soil erosion can be severe. According to a study by Sack and deLuz (2003) in Appalachia (USA), erosion rates in a riding season involve the loss of over 200 kg of soil each year – just along a 60 m trail section. Although this impact appears to vary with the different environments (coastal, Arctic-alpine, tropical and arid environment), as revealed by several studies, compaction is generally higher in wetter, poorly drained soils than in well-drained soils (Willard and Marr 1970; Burde and Renfro 1986). Soils with fine texture are more erodible than coarse or heterogeneous soils (e.g. Bryan 1977; Welch and Churchill 1986).

Trails created by off-road vehicles also mean damage to the vegetation. By this not only the size of the area exposed to soil erosion increases but, especially in sensitive areas, natural ecosystems are severely damaged or destroyed. The rate of erosion will depend on the type of vehicles and sports, the frequency of use and the type of soil in the given area (Welch and Churchill 1986).

Plate 15.4 A pathway leading to Zermatt (source: http://www.bartosik.org/scrapbook/switzerland/zermatt.htm)

Snowmobiles can have similar impacts; however, this level is estimated to be lower than those of off-road-vehicles. It must also be noted that they can be especially damaging when travelling on steep slopes or thin snowpacks (Davenport and Switalski 2006).

Over the past 50 years, the number of *dust storms* originating in the Sahara, where wind can move 65–220 million tonnes of fine sediment each year has increased tenfold. Goudie claims that up to 3 billion tonnes of dust is blown around the world annually. The number of dust storms emanating from the Sahara has increased tenfold in 50 years for which a major cause is the use of four-wheel drive vehicles replacing camels as desert vehicles (Plates 15.5 and 15.6). He blames the process of '*Toyotarization*' – a coinage reflecting the use of Toyota Land Cruisers all over in the deserts of the world – for destroying a thin crust of lichen and stones that protected vast areas of the Sahara from the wind for centuries. Four-wheel drive use, along with overgrazing and deforestation, was the major cause of the world's growing dust storm problem, the scale of which was much bigger than previously realized. Taking the whole of Sahara, and the Sahel to the South, dust volumes had increased four- to sixfold since the 1960s. The countries worst affected were Niger,

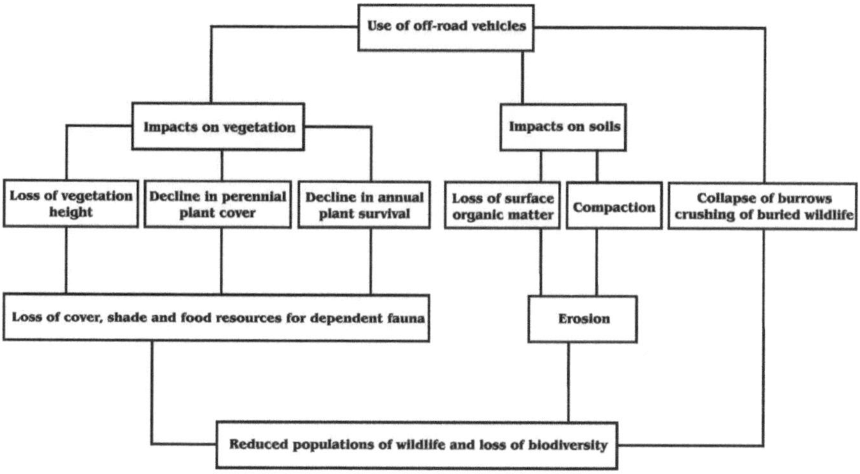

Fig. 15.8 The impacts of off-road-vehicles on the surface (by Newsome et al. 2002 from Edington and Edington 1986)

Plate 15.5 'Toyotarization' in the Sahara (source: http://www.pbase.com/pnd1/image/72911380)

Chad, Northern Nigeria, Burkina Faso and Mauritania, the research found (Goudie 2004).

Erosion impact induced by hikers, mountain bikers, horsemen and motor-cyclists were studied by Wilson and Seney (1994) in one of the forested areas of State Montana, in the United States. According to their findings, motor-cycling, horse-riding and walking, in some cases have more erosion impacts than mountain biking. Chiu and Kriwoken (2003) also claim that there is no considerable difference between the footpath erosion impacts of *hiking* and mountain biking; however, the attitude of cyclists (e.g. side-slips), a humid or even too arid and less compact surface, a younger path increases the erosion impact caused by mountain bikers,

Plate 15.6 'Toyotarization' in the Sahara (source: http://www.weirdharoldscoolparts.com/Rock%20Frog%20Little%20Sahara%202.jpg)

Plate 15.7 Widening the path for mountain biking (source: http://www.justsqueezedjuice.com/MtnBike.JPG)

without any doubt. Thurston and Reader (2001) claim that the surface-damaging impacts of mountain biking do not exceed the distance of 30 cm measured from the centre-line of the path (Plate 15.7).

Off-road vehicles are, in many cases, a tool for tourism providing access to locations where tourism or recreation takes place. However, motorized sports are enacted mainly at remote, previously inaccessible areas. Such natural areas, for many, imply the place 'far from the madding crowd' and are used for various recreational activities. However, nature is also seen and exploited in different ways and to different extents. Users involved in motorized sports represent a specific subculture for which this type of consumption has the potential to differentiate themselves from the crowd and by which they take on a different identity from the rest of the population (Holden 2000). In another opinion, it is a form of tourism representing nothing more than a form of escapism from urban life, serving little purpose beyond mindless enjoyment (Boorstin 1961). Off-road driving and racing are also regarded as activities and occasions for social interaction with friends. The amusement offered by off-road driving is now available to a greater number of participants. For them, the feeling of freedom, driving at illegal locations is of high priority regardless of its consequences being widespread as well as making the management of such impacts rather difficult.

Relaxation including mental and physical recuperation, self-realization (or self-discovery), i.e. an activity that fulfils a range of psychological needs, the new experience (adventure), gaining social enhancement as well as being fancy are among the main motivations explaining people's participation.

Despite their intensity, the impacts as a consequence of concentrated use mentioned above are localized in a narrow zone along the tracks. And it is often emphasized that by careful planning, it is possible to keep the damage at an acceptable level. For example, secondary impacts can be minimized by the maintenance of established user areas as well as roads and trails (MRG 2006). One way of achieving this is through clubs and associations from the field of tourism bearing certain code of environmental ethics.

With the advanced techniques and the growing popularity of healthy lifestyles, new tools are made available for sportsmen. Their use, however, often involves more damage. *Nordic walking* originating from Finland (Plate 15.8) in 1997, has become a popular fitness activity both in Europe and in the United States within a very short time. In order to achieve more efficient efforts and to exercise the bust, 'Nordic walkers' are helped by two sticks, furthering the disruption of vegetation and soil.

Plate 15.8 Nordic walking in Tirol (source: http://sportvakantie.eigenstart.nl)

Camping is a very popular way of living with nature both in coastal and mountainous areas. Due to the high concentration of land use, campers have significant treading and compaction impacts on the soil and vegetation on-site (Hammitt and Cole 1998). The damage caused by those sleeping rough at susceptible areas with more valuable vegetation can be even worse.

Treading is also a permanent problem associated with launching watercrafts and coastal recreation (e.g. fishing, sun-bathing) that by becoming more intensive can lead to the devastation of the vegetation and further to *coastal erosion* characteristic at coastal areas or even can quicken up this process. Madey (1994) studied the coast destruction activity of the Merced River from earlier photos, aerial photographs and by measuring channel width and coastal erosion in the most visited American national park, Yosemite. He found that recreation activities make a significant contribution to the reduction of bank stability. He described extremely high coastal erosion from campsites. Based on his research carried out in the Warren National Park (Western Australia), Smith (1998) came to the conclusion that in the surroundings of pathways leading from the camping sites to the bank of the Warren River, coastal erosion has become visually more significant due to the destruction of vegetation cover.

The role of *waves generated by the engines* of motor vessels in coastal erosion was studied by Liddle and Scorgie (1980). Such waves wash out the roots of coastal plants, leading to a decrease in coastal stability and increase in erosion.

In coastal areas attracting a large number of visitors, not only the destruction of the vegetation should be blamed for increased abrasion. Disruption of coastal sand dunes (e.g. as a result of motor-cycling, the use of sand-cruisers, pedestrians) reduces their puffer capacity also leading to denudation. Meanwhile, along the coasts of Barbados, as a consequence of various impacts from recreation, corals provide no sufficient protection, and wave erosion is intensifying (Archer 1985).

15.2.7 Damage to Landforms, Rock Collecting

Climbing is a popular sport and free-time activity of those enjoying challenge and creativity. Determined cliff-climbers break up the rocks by their bindings knocked into the hillside. This may lead to the destruction of landforms in two ways. On the one hand, it may directly cause rockfalls, and on the other, due to the scars of the surface, it makes an indirect contribution to disintegration by wind, frost and precipitation (Puczkó and Rátz 2002).

Due to their unique landforms, archaeological relics and mysterious animal life, caves are popular destinations of tourism. The number of those visiting caves has increased during the last decade (Baker and Gentry 1998). For example, the limestone cave of Derbyshire in Britain attracts 40,000 visitors each year, whereas the Baradla Cave (North-Hungary), being part of the UNESCO World Heritage from 1995 attracts ca. 200,000 (Székely 1998). The Mammoth Cave in Kentucky has about 2 million visitors annually. According to various surveys, a total of about 20 million guests are registered in about 650 visitable *show caves* worldwide in 1 year. Deliberate damages are not common, although curiosity may make visitors touch the *features* and make them detachable. This damage to stalactites, apparently insignificant, can cease their growth for good through the inhibiting wax-like material found on human skin (Puczkó and Rátz 2002).

Interested tourists often gather rocks for private collections or as a keepsake. On many occasions, even protected exposures are not favoured. Limestone blocks from the 'Great Plateau' in the Bükk National Park, North-Hungary, have been removed in great quantities for garden trim-stones (Marton-Erdős 2001). Decay of corals also occurs due to the avocation of tourists.

15.3 Conclusions

Within the system of physical environment, the transformation of any of its elements, including relief, has an impact on system functioning and can lead to remarkable transformations. The extent of such impacts on the (topographic) environment, related to recreation activities, is rather limited compared to that of other sectors like industry or agriculture. However, due to the continuously growing popularity of recreation, more attention should be paid to planning that takes the complicated relations of the environment and activities into account, with the discovery of new destinations. People looking for leisure and, in the meantime, intend to enjoy nature, will also have to become environmentally conscious visitors. The survey of the impacts of recreation activities on topography requires a multidisciplinary approach and is a great challenge to researchers. Several authors of international specialized literature have publications on the relationship of ecology and tourism (Tyler–Dangerfield 1999, Grgona 2005). Proceeding with their ideas and suggestions and taking elemental thesis of ecology and scientific approach of landscape ecology and settlement ecology as a basis, the phrase of 'tourism ecology' could be introduced. The basis, correlations and investigational territories of tourism ecology are illustrated on the following figure (Fig. 15.9). Thus, tourism ecology is a tourism development theory and practice that naturally makes possible the efficient development of rural areas building upon natural and socio-cultural resources.

Fig. 15.9 Relationships of tourism ecology (Dávid 2009)

References

Archer E (1985) Emerging environmental problems in a tourist zone: The case of Barbados. Caribb Geogr 2 (1): 45–55.
Baker A, Gentry D (1998) Environmental pressures on conserving cave speleothems: Effects of changing surface land use and increased cave tourism. J Environ Manag 53: 165–175.
Bale J (1989) Sports Geography. E & FN Spon, London
Bayfield NG (1974) Burial of vegetation by erosion material near chairlifts on Cairngorm. Biol Cons 6: 246–251
Bergsland E, Seim-Haugen O (1983) Holmenkollbakken og Holmenkollrenne. Sno og Ski 1: 97–108
Boorstin DJ (1961) The Image: A Guide to Pseudo-events in America. Athaenum, New York
Bryan RB (1977) The influence of soil properties on degradation of mountain hiking trail at Grovelsjon Geografiska. Annaler 59A: 49–65
Burde JH, Renfro JR (1986) Use impacts on the Appalachian trail pp. 138–143. In Lucas RC (ed.), Proceedings: National wilderness research conference: Current research USDA Forest service Intermountain Research station. General Technical Report INT-212 Ogden Utah USA.
Chiu L, Kriwoken L (2003) Managing recreational mountain biking in Wellington Park, Tasmania, Australia. Ann Leis Res, 6 (4): 339–361
CORINE Coastal Erosion (1998) European Communities Publication
Dale D, Weaver T (1974) Trampling effects on vegetation of the trail corridors of north rocky mountain forest. J Appl Ecol 11: 767–772
Dávid L (2009) A turizmusökológia alapvetése (The principle of tourism ecology). II. Települési Környezet Konferencia, Debreceni Egyetem Tájvédelmi és Környezetföldrajzi Tanszék, Debrecen, 55–61.
Davenport J, Switalski TA (2006) Environmental impacts of transport, related to tourism and leisure activities. In: Davenport J, Davenport JL (eds.), The Ecology of Transportation: Managing Mobility for the Environment. Springer, Dordrecht
Deluca TM, Patterson WA, Freimund WA, Cole DN (1998) Influences of lamas, horses and hikers as soil erosives from established recreation trails in Western Montana, USA. Environ Manag 22 (2): 255–262
Eckert RE Jr, Wood MK, Blackburn WH, Petersen FF (1979) Impacts of off-road vehicles on infiltration and sediment production of two desert soils. J Range Manage 32: 394–397
ECOSOC (1999) Tourism and Sustainable Development. Report of the Secretary-General, Commission on Sustainable Development, 7th session
Edington JM, Edington MA (1986) Ecology, Recreation and Tourism. Cambridge: Cambridge University Press
Erdősi F (1987) A társadalom hatása a felszínre, a vizekre és az éghajlatra Mecsek tágabb környezetében (The Impact of Society on the Surface, Waters and Climate in the Broader Environment of the Mecsek Mountains). Akadémiai Kiadó, Budapest
Goudie A (ed.) (2004) Encyclopedia of Geomorphology. Routledge, London.
Grgona J (2005) Tourism and Ecology. Annals of DAAAM & Proceedings. FindArticles.com. 20 Oct, 2009. http://findarticles.com/p/articles/mi_7105/is_2005_Annual/ai_n28321487/
Haelters J, Jacques T (1996) First National Report of Belgium to the Convention on Biological Diversity
Hammitt WE, Cole DN (1998) Wildland Recreation: Ecology and Management. John Wiley, New York
Harris J (1993) Horse riding impacts in Victoria's alpine national park. Austr Ranger 27: 14–16
Hodson M (1996) The case against golf courses. The Sunday Times
Holden A (2000) Environment and Tourism. Routledge, London
Karancsi Z (2000) Természetvédelem kontra turizmus: "Meglaposodott" brit nemzeti parkok (Nature conservation vs. tourism: 'Flattened' national parks in Great Britain). A Földgömb 18 (4): 30–37

Leung YF, Marion JL (1996) Trail degradation as influenced by environmental factors: A state of the knowledge review. J Soil Water Cons 51: 130–136
Liddle MJ (1997) Recreation Ecology: The Ecological Impact of Outdoor Recreation and Ecotourism. Chapman and Hall, London
Liddle MJ, Scorgie HRA (1980) The effects of recreation on freshwater plants and animals: A review. Biol Cons 17: 183–206
Lull HJ (1959) Soil Compaction on Forest and Rangelands. USDA Forest Service Miscellaneous Publication 768. Washington, DC
Madey MA (ed.) (1994) Analysis of bank erosion on the Merced river, Yosemite valley, Yosemite National Park, California, USA. Environ Manag 18 (2): 234–250
Marton-Erdős K (2001) Turizmus és környezet (Tourism and the Environment). Debreceni Egyetem Környezettudományi Központ, Debrecen
Mosimann T (1985) Geo-ecological impacts of ski piste construction in the Swiss Alps. App Geogr 5(1): 29–37
MRG Motorized Recreation Perspectives for TFL 38 (2006). http://www.ideal.forestry.ubc.ca/cons481/Labs/MRG..pdf
Newsome D, Moore SA, Dowling RK (2002) Natural Area Tourism – Ecology, Impacts and Management. Channel View Publications, Clevedon-Buffalo-Toronto-Sydney
Pearce D (1989) Tourist Development. Longman, New York
Phillips N (2000) A Field Experiment to Quantify the Environmental Impacts of Horse Riding in D'Entrecasteaux National Park, Western Australia. Unpublished Honours Thesis, School of Environmental Science, Murdoch University, Perth, WA, Australia
Puczkó L, Rátz T (2002) A turizmus hatásai (The Impacts of Tourism). Aula Kiadó Kft., Budapest
Ritchie B, Adair D (2002) Editorial The Growing Recognition of Sport Tourism Current Issues in Tourism 1
Romeril M (1989) Tourism and the environment: Accord or discord? Tourism Manag 10 (3): 204–208
Sack D, da Luz S (2003) Sediment flux and compaction trends on off-road vehicle (ORV) and other trails in an Appalachian forest setting. Phys Geogr 24 (6): 536–554
Smith AJ (1998) Environmental Impacts of Recreation and Tourism in Warren National Park, Western Australia and appropriate management planning Unpublished Honours Thesis, Department of Environmental Science, Murdoch University, Western Australia.
Standeven J, Knop P de (1999) Sport Tourism. Human Kinetics, USA
Strasdas W (1994) Auswirkungen neuer Freizeittrends auf die Umwelt. Forschungsbericht der Technischen Universitte Berlin. Institute für Landschafts- und Freiraumplanung. Meyer & Meyer Verlag, Aachen.
Székely K (1998) Az Aggteleki Nemzeti Park barlangjainak állapota (Status of the Caves in the Aggtelek National Park). Aggteleki Nemzeti Park
Thurot JM (1980) Capacité de charge et production touristique. CHET, Aix-en-Provence (43)
Thurston E, Reader RJ (2001) Impacts of experimentally applied mountain biking and hiking on vegetation and soil of a deciduous forest. Environ Manag 27 (3): 397–409
Tivers J (1997) Form artificiality to authenticity? The development of dry ski-slopes in England and Wales. Area 29 (4): 344–356
Türk S, Jakob E, Krämer A, Roth R (2004) Outdoor recreation activities in nature protection areas situation in Germany. In: Sievänen T, Erkkonen J, Jokimäki J, Saarinen J, Tuulentie S, Virtanen E. (eds.), Proceedings of The Second International Conference on Monitoring and Management of Visitor Flows in Recreational and Protected Areas, June 16–20, 2004. Finnish Forest Research Institute, Rovaniemi, Finland
Tyler D, Dangerfield JM (1999) Ecosystem Tourism: A resource based philosophy for ecotourism. Journal of Sustainable Tourism 7 (2): 146–158
Varga Zs (1999) A sport és a természeti környezet kapcsolata (The Relationship Between Sports and the Physical Environment). Diplomamunka, KLTE Alkalmazott Tájföldrajzi Tanszék, Debrecen

Welch DM, Churchill J (1986) Hiking Trail Conditions in Pangnirtung Pass, 1984, Baffin Islands, Canada. Parks Canada Report, Ottawa, ON, Canada

Wheat S (1995) Marking golf's card. Leisure Manag 15 (5), 26–28.

Willard BE, Marr JW (1970) Effects of human activities on alpine tundra ecosystems in Rocky Mountain National Park Colorado. Biol Cons 2: 257–265

Wilson JP, Seney JP (1994) Erosional impacts of hikers, horses, motorcycles and off-road bicycles on mountain trails in Montana. Mt Res Dev 47 (1): 77–88

Wischmeier WH, Smith DD (1978) Predicting Rainfall Losses: A Guide to Conservation Planning (Agriculture Handbook 537). Washington, DC: United States Department of Agriculture

WTO (2001) Sport and Tourism Shaping Global Culture. In: Brent Ritchie, Daryl Adair: The Growing Recognition of Sport Tourism. Current Issues in Tourism, Vol. 5 Issue 1 2002. pp. 16

Website Sources

http://www.palmsales.ca
http://www.eikongraphia.com/?p=1868
http://www.eikongraphia.com/images/TheWorldMasterplanApr07CopyrightNakheelS.jpg
http://www.westchasegc.com/sites/images/442/Westchase-Golf-Club-FL-hole-13.jpg
http://www.hnp.hu/
http://www.visegrad.hu
http://www.maridadi.com/related/feros.htm
http://www.bartosik.org/scrapbook/switzerland/zermatt.htm
http://www.pbase.com/pnd1/image/72911380
http://www.weirdharoldscoolparts.com/Rock%20Frog%20Little%20Sahara%202.jpg
http://www.justsqueezedjuice.com/MtnBike.JPG
http://sportvakantie.eigenstart.nl

Chapter 16
Impacts in Extreme Environments

Balázs Nagy

Abstract The Earth's particularly arid (hyperarid) or humid, cold or hot regions are most commonly classified as extreme physical environments. Because of their limited capacity for human settlement, anthropogenic impacts here are minimal. Being especially susceptible to external impacts, however, human presence often results in significant local alterations of landforms. In boreal areas mines operate, water reservoirs, pipelines, roads and railway lines are constructed. In extremely arid environments most human-induced geomorphological phenomena are associated with transportation, water supply, mining, military facilities and sacral activities. The arid highlands of the Himalayas, the Pamir and the Andes are populated locally even at elevations of 4000–5000 m above sea-level, and the challenging conditions created examples of extreme adaptation to and partial transformation of the physical environment. In cool high mountains traditionally terraced slopes accommodating both settlements of small extension and barley fields are established. Road cuts and rock paths represent anthropogenic landforms. Interventions of the Modern age (such as high-mountain mines, sports centres, high-mountain roads, meteorological and astronomical observatories), however, resulted in surface transformations of a new kind.

Keywords Cold environments · Antarctica · Arid environments · High mountains

16.1 Introduction

The Earth's particularly arid or humid, cold or hot regions are most commonly classified as extreme physical environments. Especially extreme cold and aridity, by which relatively large surfaces are affected, represent a great challenge to human adaptation. Due to the great differences from what is considered to be average

B. Nagy (✉)
Department of Physical Geography, Eötvös Loránd University, 1117 Budapest, Hungary
e-mail: balazs@afoldgomb.hu

J. Szabó et al. (eds.), *Anthropogenic Geomorphology*,
DOI 10.1007/978-90-481-3058-0_16,

or customary, such regions can be regarded as extreme terrestrial environments. Anomalies become more apparent when such extremities are combined – as cold and aridity in the most extreme and special cases (in high mountains and polar regions).

These spectacular extreme regions offer a limited capacity for human settlement, thus anthropogenic impacts here are minimal. With such extreme environments being especially susceptible to external impacts, human presence often results in significant topographic alterations of local scale. Human-induced geomorphic processes occur sporadically in isolated patches but can be of decisive nature.

16.2 Cold Environments

The *extremely cold areas* of the world are sparsely populated and today only a low number of population inhabits some of their regional centres. Most of the Inuits in the northern parts of North America live in winter settlements; the extreme residential environment in Siberia consists of small towns and village centres; whereas on Antarctica, human presence is restricted to container stations. Population density outside such centres and its impact on the environment therefore is rather low: human-like columns (inukshuks) of the Inuits set on stone heaps support the driving of caribou herds and hunting (as animals tend to recognise such figures as humans); Inuits and the samanist indigenous people of Siberia erected *stone mounds* at their sacred places, although the hunting–fishing and the semi-nomadic grazing life-styles do not cause any changes worth mentioning in the landforms. It is a principle on Antarctica that researchers carrying out fieldwork should not leave any traces in the area (also meaning that it is prohibited to set an open fire, or that temporary toilets and their content must be transported).

Despite this, boreal areas do not lack significant anthropogenic environmental impacts as such northern cold regions experience the operation of mines (e.g. on Svalbard); for the construction of water reservoirs, pipelines, roads and railway lines – similar to the building of homes – serious obstacles have to be surmounted. Among these the most relevant is the adaptation to the permafrost bedrock (Pinneker 1990). Soils of the *permafrost* regions thaw at summer only to a depth of some decimetres or metres; this active layer however becomes oversaturated, inaccessible and impractical. Such regions are accessible mostly in wintertime when *construction* is only possible with the soil section reframed warmed in situ. The local climate of settlements can impact the frozen bedrock causing its degradation. This can fundamentally alter geomorphic processes and water balance in periglacial areas, as a rather significant amount of water is originated here from the temporal permafrost thaw (Plate 16.1).

In cold regions *settlement* has the most remarkable topographic impact on the environment. Rather picturesque examples of this can be seen on Antarctica. Human presence only dates back to 50–60 years, there are no settlements in the conventional sense, and dwellings are represented by bases situated, in some cases, several

Plate 16.1 Storage basin between dykes near the buildings of the Arctowski Station on Antarctica (Nagy 2003)

100 km from each other, mainly on deglaciated surfaces. Life, for the some tens of people usually residing there, without any external help is unfeasible. Apart from oxygen, nearly everything has to be transported there – for many stations, even drinking water – whereas garbage should be taken away. The establishment of these bases has permanent geomorphological consequences from a number of points of view (Donachie 1993).

16.2.1 Station Construction

Station construction requires *landscaping*. Most bases are built along the coast, on bars or on wave-cut platforms. The building of storm-resistant containers, petrol tanks, warehouses on frozen ground is also associated with landscaping. As transportation-related expenses are extremely high, any waste generated by construction activities is frequently built in the bedrock.

16.2.2 Transportation

Polar environments are altered by transportation in most spectacular ways. Although ports are no longer constructed, there are three airports, established by the full levelling and compaction of moraine sediments in a zone of 1–2 km width, in Antarctica. Compared to the pristine Polar landscapes, the most visible features around stations are *wheel tracks* and *footprints*. In this extremely susceptible environment, vegetation grows rather slowly; thus its regeneration is also surprisingly slow. Footprints can remain on the mossy bedrock for years whereas caterpillar tracks remain visible for at least 30–40 years (Fiar and Nagy 2004). Wheeled and tracked vehicles are in widespread use at nearly all stations, and the tracks encompass coastal areas like abiding lines. This is due to the extremely *intensive mechanical weathering* on deglaciated surfaces; due to wind gusts, lag-sheet formation takes place within a period of few years and, even, intensive coastal chemical weathering can also cause crusting. On the rocky tundra, basically a 'desert pavement' is formed that can easily collapse under vehicles leaving traces on the surface as marked elements – even if it was passed by one vehicle once only. It is easy to imagine the degree of pressure by *tourism* concentrated on Antarctica, if a station capable to accommodate only a few persons can receive as many as 300 tourists (only for a duration of 1–2 hours) (Donachie 1993).

16.2.3 Water Supply

The water supply of stations is a rather difficult task. Meltwater has large amounts of suspended load (mainly silt), and thus requires deposition before it can be used. Moreover, its discharge is highly variable. Occasionally, supply is provided by ponds formed in the depressions of ground moraine, whereas most commonly by meltwater streams. Water running towards coastal areas is collected in *deep reservoirs* embanked by ramparts and, in many cases, underfoiled. Such lake basins can be 100 m in diameter and their considerable depths facilitate water supply during the winter. Of the rubble extracted, water-diverting dams or ramparts are constructed.

16.3 Arid Environments

In extremely arid environments, the population is concentrated mainly at oases. Most human-induced geomorphological phenomena are associated with transportation, water supply, mining, military facilities and sacral activities.

16.3.1 Transportation and Dwelling

In extremely arid regions, e.g. in the Atacama Desert of Chile, main roads follow abandoned river beds. These *wadis*, in some cases, are dry for decades, and even a concrete *road* of high quality could be constructed along the bed. Rainfall events are rather rare; the surface is predominated by the intact, distinctive landforms of previous water regimes. The desert sometimes receives waterflow, especially at intensive periods of El Niño when creeks from the Andes reach Atacama and sweep away roads.

Paved roads can, however, divert temporary branched watercourses arriving from the surrounding mountains to desert basins. Consequently, in the extremely arid basins of Iran, saline lake basins can remain without water, whereas diverted watercourses threat human settlements with flooding. In such arid basins, most of the old buildings are constructed of clay. Salt-cemented clay is an especially excellent building material. *Clay buildings* of the Incas are still seen in the Atacama Desert, and no continuous warfare and earthquakes could cause serious damage to the medieval clay fortresses Iran either. There is only one enemy they have to face: water (though early attackers were not familiar with the water cannon), and diverted watercourses easily undercut walls excellently adapted to the physical conditions.

Wheeltracks are among the most particular anthropogenic features of the micro-topography of deserts. In the troughs left by tracks passing through desert pavements and lag surfaces, crusting and deflation will start again and tracks are preserved a few centimetres deeper.

16.3.2 Water Drainage

Along the edge of Altiplano near La Paz, Bolivia, infrequent rainfalls formed extremely dissected *badlands* surfaces. As there is a fast increase in the urban population, even this gullied area is cultivated. The diversion of water from abrupt, intensive downpours seemed to be an appropriate solution to start landscaping; the watercourses diverted, however, caused the saturation of sediment to a degree that vast rotational landslides have been induced – in areas already populated.

16.3.3 Water Supply

The water supply of arid regions encircling deserts is a basic precondition of human settlement and farming. Even some hyperarid deserts unsuitable for habitation offer opportunities for water supply. The water to be utilised is in most cases not of local origin, and through draining the vital water, characteristic man-made landforms developed even further away from settlements. An old and exquisite way of water conveyance is the construction of *underground channels*. Such canal systems carrying the spring water, poor rainwater and spring meltwater of mountain ranges along the desert margins to inland regions were built from North Africa to the Middle East since ancient times. Such conducting systems, called *qanats* in Iran and *kareezes* in the Sahara, carry water originating in the mountains and straight infiltrating into debris slopes and arid debris fans (in other words, water quickly disappearing from the surface) on hillsides and foothills. In Iran the qanat system totals several ten thousands of kilometres.

For qanats *tunnels* with a gentle slope running 5–20 m deep are dug and this involves large-scale earth removal. To dig and maintain the underground water collection and drain passages, vertical pits have to be prepared, and tunnels connect the *pits* horizontally (Cholnoky, no date; Probáld 2002). The tunnels slightly slope from the mountain ranges towards the interior of the basins, while the system is ventilated through vertical shafts, from a distance seen as a series of large mole-hills (Plate 16.2). The pits generate draught, and, therefore, the water flowing underground remains cool. The earth-piles around the pitheads are gradually elevated as all deposits extracted during the cleaning of tunnels are pulled up and piled there.

Plate 16.2 Kareez pitheads in the Moroccan section of the Sahara (Nagy 1998)

Due to the significant loss by evaporation, water storage in the arid regions may require the construction of *underground reservoirs*. Around the first ancient facilities of this kind towns were erected – using the stones extracted during reservoir construction. Under the settlement of Rasafa of Roman age in the Syrian Desert, vast cisterns of three to four storeys are found, whereas of the stones extracted, a settlement with several thousands of inhabitants was built in the middle of the desert.

16.3.4 Quarrying

Wherever large amounts of *non-ferrous metals*, *building stones* (e.g. in Egypt) or *salt deposits* are found in the desert, in spite of the extreme environment, quarrying begins. As salt, in certain regions, is a valuable and fundamental raw material, food and forage stuff, it is still extracted manually and then transported by camel caravans at the hottest part of the world, i.e. in the Danakil Desert of Ethiopia. Due to unbearable heat, however, long-term extraction on a commercial scale has failed (Middleton 2004).

16.3.5 Warfare

Warfare constructions, fortresses in arid environments were adapted to the topographic and climatic endowments. Since there was no possibility to dig moats, they were strengthened most commonly by the rebuilding and fortifying the top and edge of desert mesas, thus by the building of 'citadels' (*ksars*) or fortified villages on plateaux (in Morocco and Israel). Effective defence could be also achieved by the construction of trenches, often in summit position. The oasis of Palmyra and caravan roads in the Syrian Desert are also controlled by *fortresses*, significant in size, built on a hilltop. Around the fortress with a wall height of 15 m, a *moat* of the same depth was carved into limestone (the fortress itself was built of the limestone extracted from the moat), consequently the fortress looks like if it was erected on a rock elevated in the middle of a large crater. Due to its position at the top, nothing is seen of the moat from underneath; the pinnacle of the separate hill covered by moving debris is almost impregnable (Plate 16.3).

Plate 16.3 The main defence system of the fortress of Palmyra along the Silk Road, is a deep round ditch on hilltop (Nagy 2006)

16.3.6 Sacral Activities

Certain points of the arid regions are *sacral sites* of a specific significance, and exhibit human-made landforms. Vast formations drawn by rock-piles, surface ripples or just rocks turned upside down are found on the bare slopes of the Atacama

Desert (desert rocks are different in colour in their lower sides without weathering crust, desert varnish, than on their polished side exposed to direct solar radiation). These *geoglyphs* of debated origin are – in addition to the roads running across deserts – the most characteristic anthropogenic landforms of arid regions.

With no other solution available, the followers of the Persian Zoroastrian faith – respecting the basic natural elements, i.e. earth, water, fire and air – sacrifice their dead placing them in the *Towers of Silence*, exposed to birds of prey. These large, round-shaped towers, in the arid Central Basin of Iran stand on the top of separate clusters of cliffs, hills or mounds in the vicinity of settlements. *Hilltops* are, as a result, entirely built-up by artificially supported natural walls, the towers 'reach up the sky', creating the typical elevations of desert basins.

In extremely arid environments, *permanent springs* are of great significance; a spring cave in the middle of the Central Basin of Iran being one of the most important site of pilgrimage and sanctuary of Zoroastrians. In the hollows in the hillside of the limestone range, water leaks all the time (referred to by their local name being Chak-chak, i.e. 'drop-drop'), and the karstic cavern of the sacred spring is partly *incased*, and functions as a place of worship today. With terraces carved into cliffs in its surroundings, groups of buildings provide dwellings for hundreds of people.

The worship of fire also led followers to the arid and cool summits of *volcanoes*. On the hardly climbable, debris-mantled peak of the Erciyas Volcano in Central Anatolia (3916 m), large artificial hollows can be seen in the side of the tower-like dacite pinnacles: *sanctuaries* were suitable to accommodate a number of people were carved by Zoroastrians on the severe summit (Plate 16.4).

Plate 16.4 Cave hollows carved by Zoroastrians at the 3916 m high peak of the Erciyas Volcano in Turkey (Nagy 2000)

16.4 Cold and Arid Regions

Simultaneously cold and arid regions are probably the most extreme habitats on the Earth. Mainly the plateaux of mountains and isolated valleys are classified as such (as well as some inhabited areas on Antarctica,e.g. the Dry Valleys, which have not received rain for 2 million years). The arid highlands of the Himalayas, the Pamir and the Andes are populated locally even at elevations of 4000–5000 m above sea-level, and the challenging conditions created examples of extreme adaptation to the physical environment as well as its partial transformation (Plate 16.5).

Plate 16.5 The extensive monastery group of Phuktal built in a spring-cave hollow in the Himalayas of India (Nagy 2005)

16.4.1 Dwelling

Barley is grown at a height of 4000 m in the northern arid regions of the Himalayas. This is only possible by having the *slopes terraced* and collecting topsoil gathered from larger areas. Pathways and caravan roads also run on artificial terraces and benches carved into bedrock. The most particular habitable and altered sections are the dwellings of Buddhist monk communities. The Phuktal monastery group was built in a hollow of a vast *spring cave* of a vertical cliff, the craggy valley side of one of the River Indus tributaries. This bird's-nest-like monastery constructed by broadening the miniature rock terraces, elevating piled rock-walls and the widening of foothold rock benches was, for a long time, the home of the scholar of Tibet, Alexander Csoma de Kőrös. The only caravan road of the valley also crosses this cluster of buildings through tunnels and narrow passages. The surrounding slopes, hardly accessible for humans, were terraced over the centuries, and rock paths lead along the cliffs.

16.4.2 Transportation

Road construction, however, advances slowly even in the remotest Himalayan regions. In the wake of border conflicts between India and Pakistan, military roads started to be constructed. In many cases, an annual progress of no more than 5 km is made, as roads must be carved into the walls of gorges or built on debris slopes. Completed road sections then demand permanent maintenance as unstable slopes, also jointed by explosions, constantly rip, and at times of rare rainfall events, the road is covered by thin *debris and mudflows*. (The traditional transport roads of these regions are rivers frozen in winter used by caravans or even children going to public schools.)

The so-called Pamir Road passes the Pamir Region, the marginal region of strategic importance of the Russian Empire or the former Soviet Union. This route, also accessible by motor vehicles, reaches a height above 4000 m, traverses arid and cold mountain deserts, gorges and outwash plains. Across the latter, the road passes on bulldozed gravel embankments – as effective dams against high-energy braided rivers of meltwater flow. The road opened the inland arid plateaux of the mountains (called 'pamirs') for trucks that, starting from the bank level of 4000 m, often use the arid beds of saline mountain lakes with oscillating water level. Therefore, the collapsed desert pavement formed in the vast desiccated lake bed sections preserves the *deep-cut wheel tracks* until the following flooding (i.e. for decades or centuries). There are no permanent roads in the mountain deserts of diurnal freeze–thaw alterations; patterns of wheel tracks preserved in the desert pavement are the most remarkable linear landforms of the plateaux.

To the railway line crossing the Qinghai-Tibet Plateau to Lhasa, one of the greatest challenges was the permanently frozen bedrock. In the environs of the *highest railway line of the world*, a so-called *warm permafrost* is found, i.e. soil temperature is only 1-2°C below freezing point. Railway constructions built on frozen bedrock can dislocate even following a slight warming. Therefore, frozen conditions had to be maintained here by applying a successful Chinese technology, in use since the 1980s (Brown 1984), fine-grained, insulation rock debris was spread beneath the rails, that, as a protective layer can significantly delay (or impede) the warming of subsurface layers (Permafrost warming... 2004).

16.4.3 Mining (Sulphur, Rock Salt, Ores)

On the Highlands of Bolivia (Altiplano), i.e. the cold and arid central highlands of the Andes, at similar elevations, vast salt deserts (*salars*), saline lakes and mountain hamadas are found between giant volcanoes. Open-pit *sulphur* mining takes place at extremely sparsely populated regions above 5000 m above sea level; test-drillings developed artificial geysers; on the salars, intensive *salt* mining is practiced. Locally, salt is traditionally simply scraped off from the surface and piled in heaps, but more usually ripped by axes, saline water is gathered in hollows and artificial basins.

The spatial extent of human intervention is very restricted on the salars of several thousands of square kilometres. Characteristic polygonal salt formations, however, are obliterated by the swelling and shrinking of the wetted and dried salt mass; some powdered salt can be blown away by the wind as well.

The margins of the Altiplano do not lack remarkable impacts of human intervention. The town of Potosí at 4100 m elevation with a population exceeding 150,000 inhabitants is dependent on the Cerro Rico (Mountain of Richness) rising above the town. This volcanic ruin is extremely abundant in *non-ferrous and precious ores*, still exploited – except for the use of dynamite – by long outdated technologies (Barron 2000). Despite this, typical mountain mass landforms surrounded by waste tips, with terraced slopes, dissected by mining are developed also at this elevation.

16.4.4 Science

Along the edge of Lake Titicaca plateau, extensive plug domes rise. Columnar and sheet-like lava crusts outcrop on heights dissected by erosion. The remains of the incurving lava layer of hyperfine structure run along the slopes and on the hilltops like several metres-high ridges, with narrow corridors between them. The region's Indian civilisations, Aymaras and Kechuas, claim this region to be the birthplace of their folk and culture. The crust structure of the plug domes, with minor alterations made, was used as a basic element of their calendar (Plate 16.6). Looking through the narrow corridors between the vertical lava combes only a small section of the sky can be seen. Passages are covered by horizontally laid *flagstone*s placed on *benches* carved into the combes, constraining these corridors more and letting the Sun only shine through the passages at certain time periods; moreover, the location and movements of celestial bodies were also traceable in this pipe-like 'crosshairs'.

16.5 High Mountains

Cool high mountains with diverse topography are, without any doubt, the most extreme sites for human settlement. There has to be an explanation for the presence of humans in such environments: this can be financial, a cause of power or politics as well as research interests. Certain ethnic groups, e.g. in Tibet, in the Himalayas or in the Andes could well adapt to the high-mountain living conditions and extended their dwellings to a height of 4500 m. At this level, traditionally terraced slopes accommodating settlements of small extension and barley fields are established (in the Himalayas), as well as road cuts allowing the passage of mule and yak caravans and rock paths represent human-made landforms. Interventions of the Modern age resulted in surface transformations of a new kind. High-mountain *mines* were opened (even above 5000 m elevation in Bolivia), *sports centres* were established (the world's highest ski-resort for all-year use can be found at an elevation of

Plate 16.6 Astronomical observation device formed of plug-dome crusts at Lake Titicaca (Nagy 2004)

5600 m above sea-level in Cordillera Real of Bolivia), high-mountain *roads* were constructed (crossing a pass at 5800 m in the Himalayas in India). Border conflicts and *military actions* do not avoid high mountains either: at the border regions of India, Pakistan and China, or Chile–Argentina and Bolivia bases, military roads and airports were built. Locations of the Himalayas above 6000 m, in glacial environments were also the sites for combats; Argentina and Chile maintain military bases on the ice-sheet of the Patagonian Icefield.

16.5.1 Transportation

High-mountain road construction faces challenges mainly due to *rockfalls*, *avalanches* and *frost action*. In such an environment, perfect tracks can never be found; thus these roads have to be effectively shielded from mass movement processes. In any other case the road becomes unsuitable for use. The construction of *avalanche retaining walls*, paved roads is rather expensive; thus, according to, e.g. French plans, Transfăgărăşan Road of Romania built in the 1970s can only be used in the summer and autumn months. Moreover, its section crossing a *tunnel* under the main ridge is so drenched during wintertime, hard and enduring water–ice mass is formed in it (Nagy 2008).

Road constructions can make hillside slopes unstable leading to accidents, e.g. in the late 1980s in Northern Italy where a serpentine road passing in one of the valleys of Disgrazia Hill – at its most switchbacked section – caved in with the hillside during a rockfall (Plate 16.7). A number of vehicles got stuck in the upper dead end of the valley. The vehicles remained there for years, until a new road was constructed and they could be used by anyone who carried fuel to fill their tanks.

Plate 16.7 High-mountain roads are often cut into cliff walls as in this valley of Western Tibet at 4000 m elevation (Nagy 2001)

An inundation following a downpour in the restored valley floor destroyed the broad road of marl crossing the cleaned bottom of the High Lapushnik in the Retezat Mountains of Romania. Cars marooned were taken into pieces and carried down in rucksacks, except for the bodyworks. Evidences of destructive floodings, explicitly anthropogenic in origin, are also known in mountain valleys. Although dams are now not built in the Carpathians (remains of an old dam was also destroyed by the above-mentioned flooding in the Retezat Mountains), artificially inundating steep valley sections was a frequently used method of *timber rafting*. Valleys were impounded by wooden dams, fallen trees were pulled to the valley floor and piled partly in the impounded lake, partly along the lower sections; then by opening the floodgates, timber floated down to the foothills. This method was obviously associated with extremely intensive erosion, abrasion, and valley floor degradation.

16.5.2 (Opencast) Mining

Although not being an extreme environment, the vast sulphur mine in the Kelemen (Calimani) Mountains is the largest opencast mine in the area. Exploitation began at the top of the Eastern Carpathians at 1900 m elevation. In the Calimani Negoi a 20-levelled mine, 400 m deep with a diameter exceeding 1 km has been formed (Plate 16.8). A near U-turn took place in the 1970s, in addition to what a major amount of waste material was piled in a height of 1500–1700 m. Waste material covers some of the main ridge and impounds valleys (Romania. . . 2004).

Plate 16.8 A sulphur mine hollow 400 m in depth in the caldera of the Kelemen Mountains (Nagy 2006)

16.5.3 Sport (Ski Tourism)

High mountains are the sites of mountain sport activities becoming more and more popular. The soil mechanical, surface degradation and mass movement consequences by the construction and maintenance of *ski tracks* are well known by the examples from the Alpine countries (the use of snow cannons and snow compaction by caterpillars cause *soil compaction*, and the reduced water drainage can be the cause for lateral landslides), in addition to what snowboarding off ski tracks is a typical human trigger of avalanches. Mountain shelters, helicopter platforms require *landscaping*, whereas tourists are mainly responsible for treading, which can be surprisingly severe in particularly susceptible areas visited by a large number of tourists. Debris slopes 3 km in length at the northern ice-free side of the Aconcagua, the highest summit of South America (6962 m), are ideal ski tracks: hundreds of people tread there day by day in January and February, moving a significant amount of rock, also slowly moving upward on the extremely tread serpentine road. The camps, sometimes comprising hundreds of tents on the moraine surface of the base, alter the high-mountain environment with their windbreak rock-walls (Plate 16.9).

Plate 16.9 Tourist path in an avalanche channel in the Rodnei Mountains (Nagy 2002)

16.5.4 Science (Astronomy, Meteorology)

High mountains are also the sites of scientific research. Meteorological and astronomical *observatories* (sometimes both) can be characteristic objects in extreme environments. Their construction and operation are usually associated with the transformation of the top region. In many cases, there is a cableway leading to the summit; helicopter platforms are built; the top-level is terraced or levelled. Often, the process of construction itself has the greatest impact of transformation on the landscape (by, e.g. access road construction). In other cases, the evolution of the environment is changed by the houses built.

In the case of the Carpathians, a range of examples can be seen. The observatory at the High Tatras' Lomnitz Peak is a station with typical transformation of summit also supplied by a cable way. The station at the highest peak of the Bucegi Plateau (Omul Peak) stands on a broad summit, can only be accessed on foot and is surrounded by remains of buildings and high piles of construction rubble. The meteorological station in the Rarau Mountains of the Eastern Carpathians is built near a wet spring bog, modifying runoff and water flow direction. On one of the peaks of the Cerna Hora in the Maramureş Mountains, i.e. on Pop Ivan, a four-storey high giant observatory was erected. On the summit, only accessible on foot, the observatory was built by Poland prior to World War II. Pieces of the giant copper dome weighting tonnes were carried up by horses. It could not be put into operation when the war broke out, and became a well-defendable military post, later left to decay. It is surrounded by trenches and machine-gun nests deepened into the ridges.

References

Barron A (2000) Potosi's Silver Tears. http://www.unesco.org/courier/2000_03/uk/dici/txt1.htm
Brown J (1984) US permafrost delegation to the People's Republic of China. North Eng 17 (1). http://www.uaf.edu/esm/esmpoll/perdel.pdf
Cholnoky J (no date) A sivatag (The Desert). Franklin Társulat, Budapest
Donachie SP (1993) Scientific and tourist activities. In: Rakusa-Suszczewski S (ed.), The maritime Antarctic Coastal Ecosystem of Admiralty Bay
Fiar S, Nagy B (2004) Olvadó jövő – Magyar expedíció az Antarktiszon ('Melting future' – A Hungarian Expedition on Antarctica). General Press, Budapest
Middleton N (2004) Extrém utazások (Extreme Travels). General Press, Budapest
Nagy B (2008) Keresztül a hegykoszorún – A Kárpátok magashegyi útjai (Across the Belt of Hills – Mountain Roads in the Carpathians). A Földgömb 7:36–43
Pinneker J (1990) Groundwater monitoring and management in permafrost areas. Groundwater monitoring and management, IAHS Publ. 173. http://www.cig.ensmp.fr/~iahs/redbooks/a173/iahs_173_0039.pdf
Probáld F (ed.) (2002) Afrika és a Közel-Kelet földrajza (The geography of Africa and the Middle East). ELTE Eötvös Kiadó, Budapest
Permafrost warming – a challenge to Tibetian train route (2004). http://www.innovations-report.de/html/berichte/geowissenschaften/bericht-37706.html
Romania – Mine Closure, Environment, and Socio-Economic Regeneration Project (Vol. 8 of 22) (2004) Calimani Negoiu Open Pit. The World Bank Documents and Reports. http://www.minind.ro/Pagina_noua_UMP/management_17_05/CALIMANI_EMP.pdf

Part IV
An Anthropogenic Geomorphological Synthesis

Chapter 17
Nature and Extent of Human Geomorphological Impact – A Review

Péter Rózsa

Abstract The importance of human society as a geomorphological agent is indicated by the fact that at least one-third of the Earth's continental surface is the scene of human activities. The amount of earth moved during and by different activities seems to be the most proper index to measure human impact on the Earth's surface. The different estimations suggest that anthropogenic processes are the dominant geomorphological factors. Potential human impact on the surface is basically determined by contemporary technical development level and population number. The most useful model suggested so far for quantification of potential anthropo-geomorphological impact is Nir's 'index of potential anthropic geomorphology' (I_{pag})-based 'the degree of development' and 'the degree of perception' concerning the rate of human impact and perception of the damage from anthropogenic processes, respectively. This anthropic geomorphological model can be regarded as a pioneer attempt; however, some details of the concept are debated and call for further refining and renewing modifications.

Keywords Geomorphological activities · Socio-economic factors · Population growth · Anthropo-geomorphological processes (AGP) · Index of potential anthropogenic geomorphology (I)

17.1 Introduction

Although it has numerous forerunners, anthropogenic geomorphology is a relatively new subdiscipline. On the one hand, it is true that G.P. Marsh's pioneer work (*Man and Nature, or Physical Geography as Modified by Human Action, New York, 1864*) and principal books of R.L. Sherlock (*Man as a Geological Agent – an Account*

P. Rózsa (✉)
Department of Mineralogy and Geology, University of Debrecen, Egyetem tér 1, 4032 Debrecen, Hungary
e-mail: rozsap@puma.unideb.hu

J. Szabó et al. (eds.), *Anthropogenic Geomorphology*,
DOI 10.1007/978-90-481-3058-0_17,

of his Action on Inanimate Nature, London, 1922; Man's Influence on the Earth, London, 1931) were published almost 150 and 70–80 years ago, respectively; on the other hand, however, geomorphological studies began to focus on anthropogenic landforms and processes as late as the 1960s. Due to ever intensifying research, extensive knowledge has accumulated and special aspects of this subdiscipline have also been formed during the past 50 years.

The *geomorphological approach* aims at studying the magnitude and rate of anthropogenic geomorphological processes. *A socio-economic approach* focuses on the economic and social influence on the dynamism of human geomorphological activity. Changes of human impact on geomorphological processes over a time interval are studied by the *historic approach*. The *planner's approach* unifies the aspects mentioned above (Nir 1983).

Reviewing anthropogenic geomorphology, all of the above aspects have to be considered. It has to be noted, however, that there is no widely accepted anthropogenic geomorphological synthesis of this kind, although some remarkable publications have appeared on this topic. The aim of this chapter is to provide quantitative and qualitative evaluation of human impacts on the Earth's surface by a critical review of principal publications.

17.2 Estimating the Rate of Anthropogenic Geomorphological Processes

The importance of humankind as a geomorphological agent is indicated by the fact that at least one-third of the Earth's continental surface of 149 million km^2 can be regarded as a scene of direct or indirect anthropogenic geomorphological activity. At the turn of the millennium, arable land and plantations covered 15 million km^2, grazing land was 35 million km^2 and built-up areas occupied 2 million km^2 (Loh and Wackernagel 2004). Moreover, a considerable part of forests (38 million km^2) has also undergone intensive human transformation.

The areal extension of human activities, however, is not suitable in itself for either a quantitative or a qualitative evaluation of anthropogenic geomorphological impact on the Earth's surface because the influence of economic activities related to each land-use type may extremely differ from each other. Built-up areas, covering less than 2% of the continents, for instance, suffer the most intensive surface modifying activity from humans. Comparing, for example, 'natural fluvial' to 'anthropogenic' erosion for the United States (Fig. 17.1), it can be noticed that human activities move much more earth; moreover, there are significant geographical differences between the two types of erosion: while fluvial erosion is in close connection with relief, the most intensive anthropogenic impact occurs in highly urbanized and in mining areas (Hooke 1999).

The quantity of the earth moved during and by various human activities seems to be the most appropriate index for the estimation of the extent of human impact on the Earth's surface. This estimation, however, is not simple at all. On the one hand, some

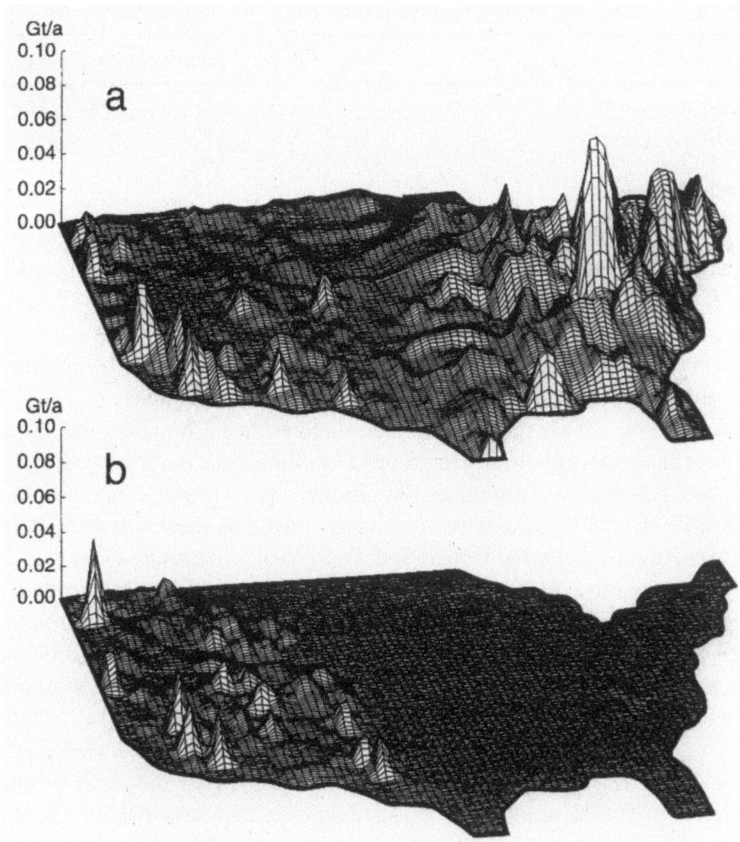

Fig. 17.1 Spatial distribution of the earth moved per year by human activities (**a**) and fluvial erosion (**b**) in the USA (Hooke 1999)

human activities (forest clearing, tillage, grazing, etc.) characteristically alter topography in an indirect way, i.e. by modifying 'natural' erosion processes; on the other hand, there are no correct statistical data concerning the amount of the earth moved by direct anthropogenic geomorphological activities (urban construction, road and railway constructions, mining, etc.). Moreover, the different estimations for the efficiency of anthropogenic and natural geomorphological processes may differ from each other by orders of magnitude.

A summary estimation of human geomorphological impacts for the 1970s was published by Nir (1983). By a critical review of statistical and literature data, he concluded that, on the basis of the anthropogenic erosion rate, agriculture could be regarded as the most significant of Earth's surface modifying human activity (Table 17.1). To demonstrate the shocking mass of 173 billion tons, he mentioned that the River Huang He, one of the muddiest rivers in the world, carries about

Table 17.1 Rates of anthropo-geomorphological processes (based on data by Nir 1983)

Human activity	Rate of erosion (billion t/yr)
Forest clearing	1
Grazing	50
Land tillage	106
Mining and quarrying	15
Road, railway construction and urban development	1
Total	173

1.5 billion tons of silt per year, i.e. the estimated quantity eroded by anthropogenic geomorphological activity is more than one hundred times higher.

Hooke (1994, 2000) applied another method for estimating the efficacy of anthropogenic geomorphology. He grouped building construction, road and railway constructions as well as mining as 'intentional' anthropogenic geomorphological activities. To estimate the quantity of the earth moved by these activities worldwide, he used statistical data for the United States as a basis. According to his calculations, the amount of earth annually moved is 0.8 billion tons for building construction, 3.8 billion tons for mining and 3 billion tons for road and railway constructions.

Hooke used two parameters to determine the quantity of annually moved earth for the world from US data. He assumed that intentional human geomorphological activity in a given country is directly related to its Gross National Product (GNP) or to energy consumption. (Gross National Product, or GNP, is the value of all the goods and services produced in a country's economy, plus the value of the goods and services imported, minus the goods and services exported including profits from capital held abroad.) Since, in 1991 the GNP of the world was four times higher than that of the United States, the total quantity of the earth moved by intentional human geomorphological impact was estimated to be about 30 billion tons for that year. In the same year the United States shared 21.7% of total energy consumption of the world; considering this proportion a total value of 35 billion tons can be calculated (Hooke 1994).

Hooke regards agriculture as an unintentional anthropogenic geomorphological activity. (Although Hooke's classification does not follow the generally accepted and used terminology (see for example Szabó 1993), it can be related to that. His intentional human geomorphological intervention corresponds to mining, industrial, urban construction, road and railway constructions and, partly, riverbed and shore management activities, while agriculture and forestry are regarded as unintentional ones. Therefore, the terms 'intentional' and 'unintentional' refer to the fact that human modification of the Earth's surface is direct and deliberate in the first case, and it is basically (but not exclusively!) indirect and unwilling in the second one.) His starting point is that today the area of cultivated and grazing land per capita is 0.3 and 0.6 ha, respectively (Hooke 2000). With 6 billion as the world's population, the loss of material due to cultivation and grazing would be 63 and 36 billion tons per year, respectively. Consequently, the total amount of earth moved annually by different human activities is roughly estimated at ca. 130 billion tons. On this basis, it is claimed that anthropogenic processes are the predominant modifying factors of the Earth surface (Table 17.2).

Although the estimations by Nir and Hooke are of similar order, there are remarkable differences between them regarding both the total sum and the different human geomorphological activities. These differences result from several factors. First, it has to be considered that Nir's estimation refers to the erosion rate caused by human action, while Hooke's calculation represents the amount of the earth moved by human activities. Moreover, the authors applied different approaches: Nir principally used United Nations statistical data for the whole world, while Hooke tried to calculate the amount of human geomorphological activity for the world by extrapolating statistical data for the United States. It should be also taken into consideration that estimation of geomorphological influence of unintentional human activities (ploughing, grazing, etc.) is quite difficult because of our insufficient knowledge on the 'natural background'. For instance, estimations by different authors on the erosion rate due to grazing range from 10 to 25,000 kg/ha/year (Nir 1983). Undoubtedly, however, rural activities still seem to be the most significant anthropogenic geomorphological factors. Finally, it also has to be taken into account that Nir's data refer to the mid-1970s, while Hooke's estimation concern the early 1990s. The more than double value for intentional geomorphological activities (urban development, road and railway construction, mining) for the early 1990s can be partly explained by the expansion of these activities over the past 15 years.

Table 17.2 Summary of the estimated rates of anthropogenic and natural geomorphological activities (after Hooke 1994, 2000 and Haff 2003)

Geomorphological factor	Earth moved (billion t/yr)
Man	
Intentional based on GNP	30
Intentional based on energy consumption	35
Unintentional (agriculture)	99
Total anthropogenic	129–134
Rivers	
Long-distance sediment transfer	14
Meandering	39
Glaciers	4
Slope processes	1
Wave action	1
Wind	1
Mountain building	
Continental	14
Oceanic	30
Deep ocean sedimentation rates	7
Total natural	111

17.3 Socio-Economic Factors in Anthropogenic Geomorphology

Although human modification of the Earth's surface is of the same age as humankind itself, it would be a little exaggeration to say that human history is automatically regarded that of anthropogenic geomorphology. Potential human impact on the environment (and on the surface at the same time) is basically determined by two factors: *technical development level* and *population number* of the given age.

17.3.1 Technical Progress

Until the early Holocene (12,000 years BP), humans used wood, bone and chipped flint implements and basically followed a hunting-gathering course in life; however, the production of corn, peas and lentils as well as the domestication of some wild animals also began. This era is called *the first agricultural revolution*. The first agricultural civilizations were formed more or less at the same time but independently from each other in the area of the so-called 'Fertile Crescent' (Mesopotamia, Near East, Nile Valley), East Asia, Mexico and Peru. According to our recent knowledge, this phenomenon was the result of the shift of the climatic zones in the late Pleistocene because of which areas of the formerly temperate zone became more arid (semi-arid, Mediterranean, etc.).

The general use of *irrigation* 5,000 years ago resulted in a quantitative change in agriculture since in the hot arid and semi-arid areas it made several harvests possible in a year. Two fundamental technical inventions, i.e. of the *plough* and the *wheel*, too happened at this time. Using the plough large areas could be cultivated in an intensive way, and the application of wheel allowed transportation of more voluminous material to a longer distance.

The smelting of ores and *the processing of metals* started a new chapter in the history of human impacts on the environment. Copper was produced for the first time in Armenia, Iran and Thailand approximately 7,000–8,000 years BP. Bronze (an alloy of copper and tin) was produced some thousands of years later. In the Near East iron was known as early as 5,000 years BP; however, iron smelting became to be generally known as late as 1200–1000 BC.

Riverine civilizations formed in the Nile Valley, Mesopotamia as well as along the Huang He and the Indus Rivers about 4000 BC. The building and maintenance of irrigation canals required organization, the division of labour and a central control, which necessarily led to the formation of a slave society. The first urban settlements and urban communities were formed in Mesopotamia about 3000–5000 BC. The process of *urbanization*, the 'urban revolution', successively spread in the Near East, the Nile Valley, ancient Greece and then over almost the whole Mediterranean. At the same time, remarkable urban civilizations formed in India and China as well. Progress of ancient urbanization culminated in the Roman Empire. In the great age of the Empire, approximately 30 settlements had a population of more than 100,000 people; in the 3rd century, the number of inhabitants of Rome reached 1.5

or 2 million. A network of several thousand towns was connected by a developed road-system. The fall of the Roman Empire, however, retarded European economy and urbanization for centuries. This time Byzantium, the Arab Caliphate and China became the most developed and prospering territories in Eurasia.

After a stagnation of 500 years due to *agricultural technical innovations* as well as a more intensive utilization of *water and wind power*, an economic prosperity began in Europe as well and, as a consequence, a new wave of urban development commenced. This progress was retarded by the disastrous plague epidemic, which broke out in 1347. Before the epidemic, the population of Europe was 74 million; however, 50 years later approximately 50 million people lived in the continent; even a further 100 years was not enough to retrieve the population loss (Livi-Bacci 1992). With declining population, however, the prices of agricultural products dropped and wages increased. This situation stimulated industrial production, trade and the spreading of technical innovations such as the sluice, dredging machine, hydraulic structures, etc.

Social and economic circumstances as well as the human activities of geomorphological significance were dramatically changed by the *industrial revolution*, which began in England in the 18th century and spread over Europe and North America by the end of the 19th century. Due to the use of *coal* and the utilization of *steam power* (Watt's steam-engine – 1769, Fulton's steamer – 1807, Stephenson's locomotive – 1825), the introduction of other technical innovations (e.g. Cartwright's power-loom – 1787) and the exploitation of new raw materials, a *mechanized large-scale industry* replaced the predominantly agricultural and handicraft-based economy. The application of new capital equipment and the implementation of new work organization resulted in dramatically increasing productivity. Sudden progress in mining, transport and agriculture also ensued. The next stage of industrial development starting in the 1870s is called as *second industrial revolution*. The use of alternating current and the widespread use of *electricity* again transformed economy and everyday life. The invention of the *combustion engine* made the production of machines of higher efficiency possible, dramatically increased the demand for petroleum, radically transformed transportation and gave a new impetus to chemical industry. The development of electronics and the utilization of nuclear power from the middle of the 20th century is the *third industrial revolution* (Table 17.3).

17.3.2 Population Growth

The population before the *Upper Paleolithic* (35,000–30,000 BC) can be estimated at *some hundred thousands* and the annual rate of growth did not reach 0.1 %; therefore, doubling the population required 8,000 or 9,000 years. About 10,000 BC, approximately 6 million people lived on the Earth. During the next 10,000 years, the annual rate of growth reached 0.4 % and the population could have reached 250 million. The rate of growth, besides periods of decline, increased steadily over the

Table 17.3 The main social–economic eras (after Simmons 1993 and Goudie and Viles 2003)

Age		Time zone	Principal innovation	Energy source	Environmental impact
Hunting-gathering		Up to 10000 BC	Beginning of tool production	Human muscle	Local and short-term
Agricultural	*Early agriculture*	Up to 5000 BC	Cultivation, domestication	Human and animal muscle, wood, wind and water power	
	Riverine civilizations	Up to 500 BC	Irrigation, use of metals, spread of plough and wheel		Local, longer-term
	Agricultural empires	Up to the 1750s	Terracing, road network, utilization of wind and water power		
Industrial	*First industrial revolution*	Up to the 1870s	Spread of stem engines, industrialization	Coal	Regional, permanent
	Second industrial revolution	Up to the 1950s	Steel making, railway network, utilization of electricity, combustion engine	Coal, petroleum	
	Third industrial revolution	From the 1950s	Plastics, electronics, utilization of nuclear power, computerization	Petroleum, natural gas, nuclear fuels	Global, permanent and perhaps irreversible

following 17 or 18 centuries and world population exceeded *three-quarters billion* by the middle of the 18th century, i.e. at the beginning of the industrial revolution. From then to the middle of the 20th century, the rate of population growth was almost ten times higher (it reached 6 % corresponding to a doubling time of 116 years) and the population exceeded 2.5 billion. In the second half of the last century, the rate of population growth tripled and world population has doubled in 40 years. At the millennium, as many as 6 billion people lived on the Earth (Table 17.4).

Table 17.4 Changes in world population from 10,000 BC to 1990 (after Livi-Bacci 1992)

	10,000 BC	0	1,750	1,950	1,990
Population (millions)	6	252	771	2,530	5,292
Annual rate of growth (%)	0.08	0.38	0.64	5.96	18.45
Doubling time (years)	8,369	1854	1,083	116	38

Over the last two and a half centuries, the concentration of population has also changed dramatically. The spread of industrial revolution generated an *urban explosion process* that still continues. Technical innovations and an increase of the efficiency of agricultural production, as mentioned above, resulted in dramatic population growth. Because of the vast excess of labour in rural areas, large-scale migration to urban settlements began. At the end of the 18th century, only 3% of the world population were town-dwellers; 100 years later this ratio increased to 13.6%, in 1950 it reached 28.2% and in 2007 the world urbanization ratio reached 50%.

Population growth, on an arithmetic scale, is divided into two main stages. The first one, up to the industrial revolution, is characterized by a slow and steady increase; the second stage beginning in the era of the industrial revolution shows an exponential growth (Fig. 17.2A). A logarithmic scale, however, demonstrates three periods in the history of world population. The first period lasted from the appearance of human race to the end of Paleolithic, the second one lasted from the Neolithic to the industrial revolution and the third began during the industrial revolution and still continues (Fig. 17.2B). It must be noted that these figures are quite simplified since demographic increase is not constant within the given periods; there are longer or shorter cycles of stagnation or even decline because of, for example, disastrous wars or epidemics (Fig. 17.2C).

The three periods of population growth coincide with the three main technical–cultural eras of humankind (hunting-gathering–agricultural–industrial). The relationship between demography and economic development is obvious but complex. On the one hand, increasing population may act as an obstacle to economic development because population growth may create its own limit. In the first main period, the volume of biomass useable for food and heating and in the second one, the extension of arable land as well as the quantity of vegetable, animal, water and wind power represented this limit. At present, the limitation may be constituted by the environmental impact of technical progress (Livi-Bacci 1992). On the other hand, however, demographic pressure may enforce a more intensive production as

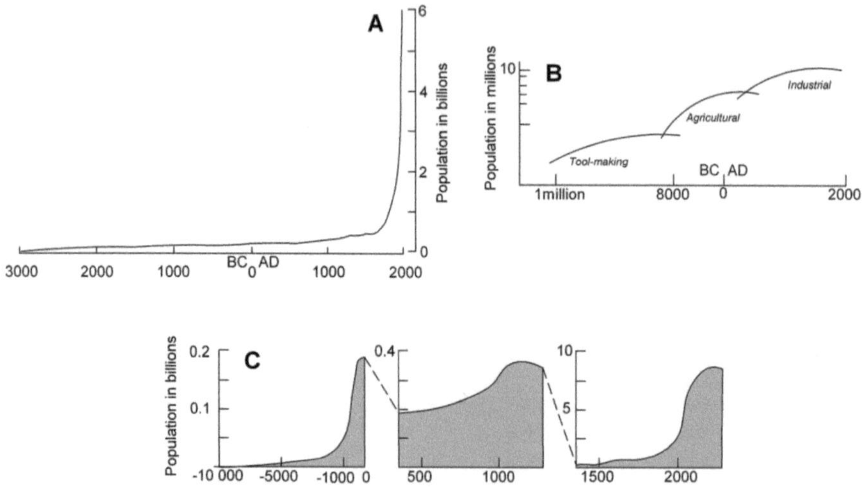

Fig. 17.2 Human population growth by arithmetic–exponential (**a**), logarithmic–logistic (**b**) and arithmetic–logistic (**c**) interpretation (after Livi-Bacci 1999; Thomlinson 1976 and Whitmore et al. 1990)

well as the invention and spreading of new technologies. Technical progress may initiate economic development, which may make a more rapid population growth possible and may create a demand for higher standards of living. This means that technical progress and population growth, the two main social factors determining human geomorphological activity, mutually intensify their influence. Moreover, the increasing demand for higher and higher standards of living may also generate further human intervention into natural geomorphic processes.

17.4 An Historic Approach to Anthropogenic Geomorphology

Hooke (2000) provided a quantitative estimation of history of human geomorphological impact. Naturally, the factual basis of his calculation can be debated since he had to rely on assumptions and premises. It is remarkable, however, that his estimation is in accordance with the general pattern, i.e. the three main technical–cultural eras of humankind identified by socio-economic factors.

As a starting point, on the basis of archaeological and historical data, Hooke estimated the mass of material moved by humans in the cases of certain relatively advanced societies in the past (Table 17.5), and plotted the data in a diagram. As the curve reflects the earth-moving activities of the most advanced cultures at a given time, it is logical that the worldwide average amount of the earth moved by humans should be lower than these data. To the worldwide average he scaled the curve by multiplying each point on it by the ratio of the per capita estimate for the world (6 tons per year) to that for the United States (31 tons per year). In his opinion, the curve obtained by this re-estimation may show the pattern of the intentional geomorphological activities by humans during the past 4,500 years (Fig. 17.3).

Table 17.5 Estimated amount of earth moved in some advanced civilizations (after Hooke 2000)

Time	Civilization	Earth moved (million tons)	Time-span (year)	Earth moved per capita (kg/yr/person)
2,500 BC	Egypt, pyramid of Cheops	6.3	20	625
0	Rome	2,330.0	800	3,495
100	Rome	290	200	1735
600	Copán, Mexico	5.3	400	665
1,400	Easter Island	1.0	600	260
1,650	London	0.9	100	3,365
1,750	London	2.0	100	4,040
1,825	London	13.0	50	12,860
Today	United States	7,600.0	1	31,000
Today	Worldwide	35,000.0	1	6,000

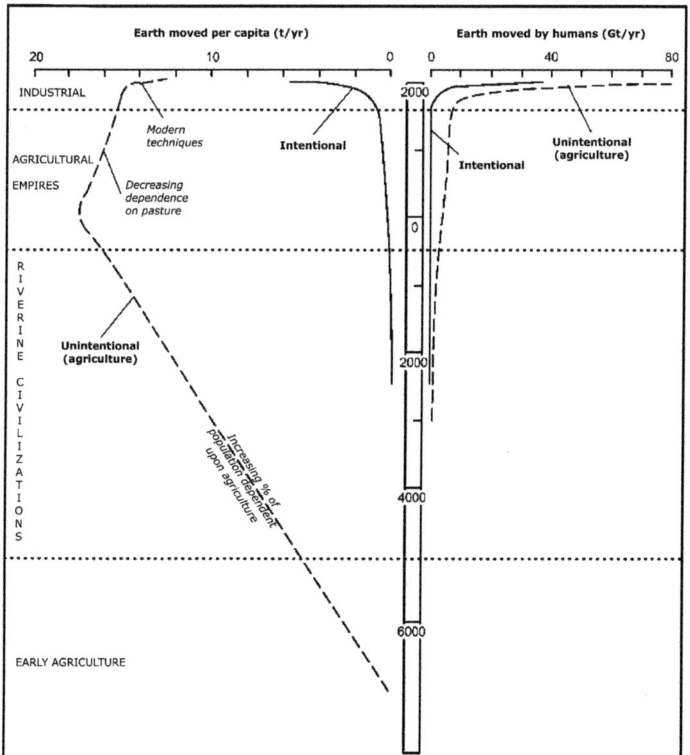

Fig. 17.3 The amount of earth moved annually per capita by human intervention from 7000 BC to the present (on the left side); and the total amount of earth moved annually by human geomorphological activities during the past 5,000 years (on the right side) (compiled by Rózsa 2007 after Hooke 2000)

To estimate the geomorphological impact of agriculture in the past, Hooke assumed that population dependent upon agriculture increased linearly from 9000 BP (i.e. from the beginning of the agricultural technical–cultural era of humankind); about 2000 BP, it reached 100% and has remained at this level. It should be emphasized that the expression of 'population dependent upon agriculture' does not refer to population following agricultural activity. This expression concerns the fact that since that time, food supply for humans has been practically provided by tillage and animal husbandry and role of fishing, hunting and gathering became rather subordinate; therefore, the food supply has been dependent on agricultural production. According to his assumption, the annual sediment loss from tillage and pasturing increased linearly during that 7,000 year period.

About 2000 BP, when riverine civilizations flourished and agricultural empires emerged, the amount of the earth moved per capita by agricultural activities began to decrease gradually. This is due to the fact that pastures became less and less important for the food supply of world population; moreover, a land of a given extent provided food for an increasing number of people; technical innovations such as iron ploughs, irrigation and others were spreading. Due to the dramatic increase in agricultural productivity and the implementation of modern soil conservation practices, the relative importance of anthropogenic geomorphological processes by agricultural activities has shown a more precipitous decline since the industrial revolution, particularly, over the last 50 years.

As indicated in Fig. 17.3, the impact of intentional anthropogenic geomorphological activities (mining, building and road construction, etc.) increased slowly until the industrial revolution and their geomorphological importance was subordinate to agriculture. After the industrial revolution, this pattern changed fundamentally: while the amount of earth moved per capita by agriculture has dropped considerably, intentional anthropogenic geomorphological activities are steadily gaining in importance.

The total amount of earth moved intentionally and unintentionally can be simply estimated by multiplying the per capita values by the population in the past (Fig. 17.3). The figure shows that geomorphological impact of humans increased slowly and gradually until the industrial revolution, and this increase was dominantly the result of the increasing unintentional (agricultural) activity. Industrial revolution drastically transformed this pattern. The most important change is that linear increase of amount of earth moved by humans turned exponential. According to Hooke's estimation, the total earth moved in the past 500 years would be sufficient to build a mountain range 4000 m high, 40 km wide and 100 km long. Moreover, at the current rate of increase, its length could be doubled in the next 100 years.

Another significant change is that, although there is a huge increase in geomorphological impact of both intentional and unintentional activities, intentional impacts have become more important. As a consequence, humans are increasingly capable not only of modifying geomorphological processes but also visually transforming the Earth's surface.

17.5 Estimating Potential Anthropogenic Geomorphological Impact

It could seem to be obvious that the amount of earth moved per capita by human activities should be regarded as a basic parameter for the quantification of anthropogenic geomorphological impact. As it was mentioned, however, this method is hardly applicable because of the lack of detailed statistical data as well as considerable variation in estimations.

Human environmental impact is generally characterized by the equation of Erlich and Erlich (1990):

$$I = P \times A \times T \tag{17.1}$$

where I is environmental impact, P is population, A is affluence per capita and T is a technology factor. Some authors (e.g. Haff 2003) suggest applying this equation to relate human factor in geomorphological processes. This assumption seems to be logical since, regarding the social–economic side of anthropogenic geomorphology, human geomorphological impact is principally controlled by population size and economic and technological development. However, this equation can be scarcely used to quantify anthropogenic geomorphological impact because there is no parameter to represent a direct and specific connection between technological development and general geomorphological ability of humans.

The most useable model suggested so far for the quantification of potential anthropogenic geomorphological impact was formulated by Nir (1983). His '*index of potential anthropogenic geomorphology*' is based on two parameters, namely '*the degree of development*' and '*the degree of perception*'. The former reflects the rate of human impact, while the latter concerns the perception of the harm of anthropogenic geomorphological processes, i.e. the extent of combating erosion caused by human intervention.

Interpreting the relationship between the degree of development and that of perception, Nir hierarchically and chronologically ordered different human activities according to their geomorphological impact and presumes that each degree contains all the earlier ones. On the basis of the rate of *anthropogenic geomorphological processes* (AGP), he identifies five different degrees on the scale ranging from 0 (AGP non-exist) to 1 (disastrous AGP). The correlation between anthropogenic geomorphological processes and the degree of development is shown by Fig. 17.4. The three curves in the figure reflect the three main possibilities for the impact of the degree of development on the rate of the AGP. Nir assumes that every curve starts and ends at the same points.

To determine the rate of the degree of perception, a similar train of thought is followed. Therefore, Nir assumes that each degree contains all the earlier, less efficient combating erosion. Figure 17.5 shows the scheme of correlation between anthropogenic geomorphological processes and the degree of perception. As the figure shows, perception may happen at low, gradual and accelerated rates, and according to Nir's assumption, the three curves begin and end at identical points in this case,

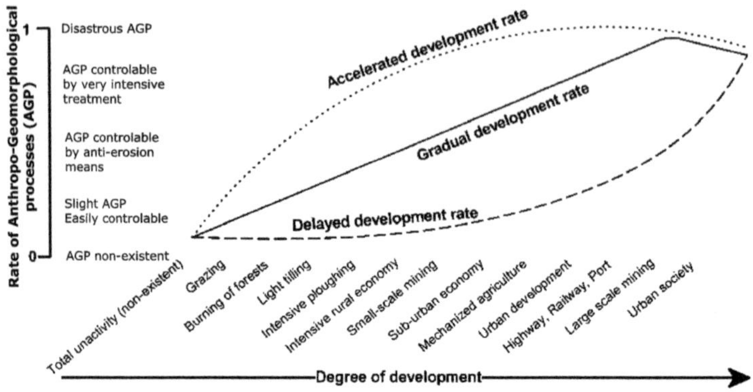

Fig. 17.4 Correlation between the degree of development and anthropogenic geomorphological processes (Nir 1983)

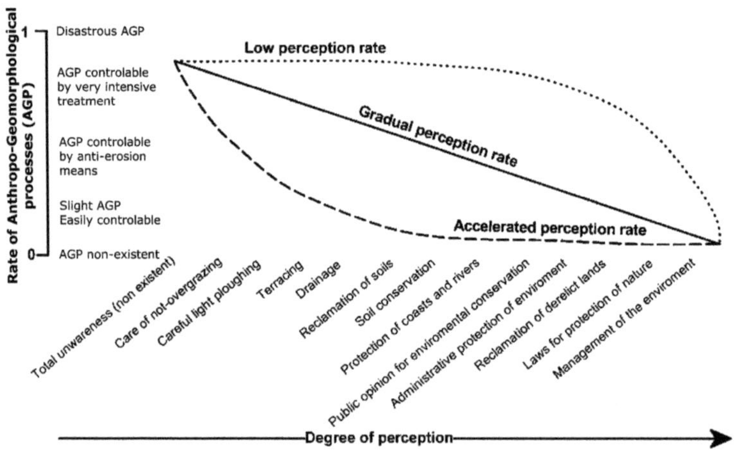

Fig. 17.5 Correlation between degree of perception and anthropogenic geomorphological processes (Nir 1983)

too. Generally speaking, it can be stated that three kinds of relationship may occur between the degree of development (DD) and the degree of perception (DP):

1. when DD is higher than DP, high AGP rates occur indicating a need for severe to moderate anti-erosive measures;
2. when both DD and DP are similar, moderate AGP rates occur and there is no urgent need for corrective measures;
3. when DD is lower than DP, low AGP rates occur.

In Nir's opinion, the degree of development can be expressed by the *percentage of urban population* (UP), while the degree of (lack of) perception can be expressed by the *percentage of illiteracy* (DI) since illiteracy may indicate education, and education is a necessary condition for forming a public degree of perception. Therefore, he proposes to define the rate of anthropogenic geomorphological processes (AGP) as the average value of these two variables, as follows:

$$\mathrm{AGP} = (\mathrm{UP} + \mathrm{DI})/2 \tag{17.2}$$

In his model, Nir intended to consider physical conditions that have an influence on geomorphological processes induced by human activities. For this reason, he also reckoned with the influence of *relief* and *climate*, the two principal physical factors modifying geomorphological processes. As a result, his '*index of potential anthropogenic geomorphology*' (I) is formulated as

$$I = \frac{\mathrm{UP} + \mathrm{DI}}{2} \cdot \frac{1}{100} \cdot (K_\mathrm{c} + K_\mathrm{r}) \tag{17.3}$$

where UP is the percentage of urban population, DI is the percentage of illiteracy and K_c and K_r are constants reflecting climatic and relief conditions, respectively. Values of constants may range from 0.4 to 0.8 and from 0.2 to 0.8, respectively (Tables 17.6 and 17.7). Higher K_c values (0.6 and 0.8) are reasoned by the significant geomorphic impact of high amounts of precipitation for equatorial and monsoon-savannah climates; that of aeolian erosion for arid and semi-arid climate; and that of gelisolifluction for cold climate. The relative values of K_r constant obviously correspond to the dissection and slope condition of the main relief categories. The multiplication by 1/100 serves to express the value of (UP + DI)/2 by a value between 0 and 1; in general, index values also range from 0 to 1.

Nir proposed this index to be a *parameter*, which indicates *how harmful potential anthropogenic geomorphological processes are in a given country*. In his opinion,

Table 17.6 Values of K_c constant for calculating I (according to Nir 1983) (Classification after Köppen)

Equatorial climate (Af)	0.6
Monsoon-savannah climate (Aw)	0.8
Arid and semi-arid climate (B)	0.6
Temperate climate (C)	0.4
Cold climate (D)	0.6
Arctic climate (E)	0.4

Table 17.7 Values of K_r constant for calculating I (according to Nir 1983)

Plains	0.2
Hills	0.4
Plateaux	0.5
Medium-high mountains	0.6
High (Alpine) mountains	0.8

if $I < 0.30$, human geomorphological activities represent a low hazard; if $0.30 \leq I \leq 0.49$, a hazard not negligible, with some conservation measures required; if $I \geq 0.50$, anthropogenic geomorphological processes may cause considerable damage; therefore, urgent and efficient measures are needed.

It is an advantage of Nir's index that the required data can be obtained for most countries. Moreover, by using prediction for urban population and illiteracy in the future, potential anthropogenic geomorphological processes can be also forecasted. Table 17.8 listed the values of I for 37 countries for the 1970s, the millennium

Table 17.8 Values of I for 33 selected countries in the years 1970, 2000 and 2015

	I		
Country	1970[a]	2000[b]	2015[b]
Algeria	0.76	0.55	0.51
Australia	0.35	0.37	0.38
Brazil	0.46	0.48	0.48
Botswana	0.31	0.30	0.30
Bulgaria	0.33	0.36	0.37
Canada	0.43	0.45	0.47
Chile	0.48	0.50	0.51
Greece	0.40	0.32	0.33
Hungary	0.16	0.20	0.21
India	0.57	0.46	0.42
Indonesia	0.36	0.35	0.39
Iran	0.60	0.49	0.49
Israel	0.50	0.48	0.47
Japan	0.30	0.34	0.35
Malawi	0.48	0.31	0.28
Malaysia	0.46	0.46	0.46
Mexico	0.48	0.46	0.46
Morocco	0.70	0.65	0.61
Nepal	0.70	0.55	0.47
New Zeeland	0.42	0.44	0.44
Panama	0.43	0.39	0.40
Philippines	0.33	0.43	0.47
Poland	0.17	0.19	0.19
South Korea	0.27	0.37	0.38
Switzerland	0.34	0.41	0.41
Syria	0.52	0.34	0.31
Tanzania	0.50	0.39	0.38
Thailand	0.17	0.18	0.20
Tunisia	0.58	0.46	0.42
Turkey	0.47	0.40	0.40
United Kingdom	0.31	0.36	0.36
United States	0.34	0.36	0.38
Zambia	0.57	0.37	0.34

[a]Data source: Nir (1983)

[b]Calculated based on data published in hdr.undp.org and www.uis.unesco.org

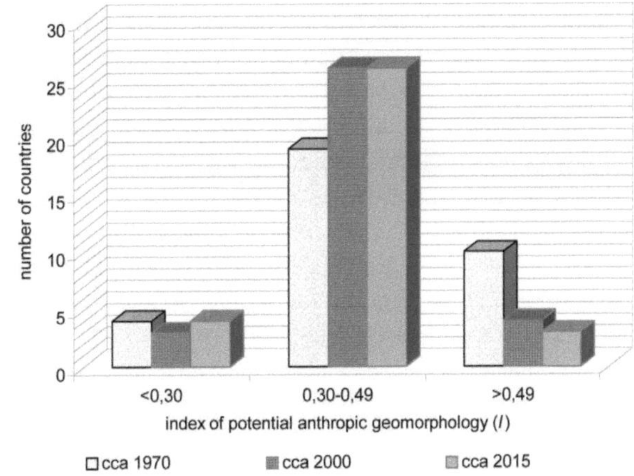

Fig. 17.6 Values of index of potential anthropogenic geomorphology (*I*) for 33 countries in 1970, 2000 and 2015 (see Table 17.8 for the data used) (by Rózsa 2007)

and 2015. Distribution of the values of *I* grouped according to the three main risk categories (Fig. 17.6) indicates that in the 1970s there were only four countries (Hungary, Poland, South Korea and Thailand) where potential anthropogenic geomorphological processes present a lower hazard; however, the high value of *I* suggested very high hazards for ten countries (Algeria, India, Iran, Israel, Morocco, Nepal, Syria, Tanzania, Tunisia, Zambia). For the other 19 countries, the medium value of the index indicated moderate hazards.

The picture seems to have improved by the millennium: the values of I suggest that anthropogenic geomorphological processes represent high hazards for only four of 33 countries (Algeria, Chile, Morocco and Nepal). Simultaneously, the number of countries of moderately dangerous human geomorphological impact increased from 19 to 26 (almost 80%). For three countries (Hungary, Poland and Thailand) anthropogenic geomorphological processes continued to be on a less dangerous level. According to the prediction for 2015, basically, the present pattern will not change.

On the other hand, however, some details of the concept can be debated. Regarding K_c and K_r constants, Butzer (1984), referring observational evidence from Jansen and Painter (1974), argues that relief heavily outweighs the influence of precipitation seasonality and intensity. Moreover, the percentage of urban population as a parameter of the level of development may be questionable. It is undoubtedly true that there are conditions of obtaining urban rank (e.g. population, offices, services, spatial attraction, urban building methods, etc.), which reflect the level of development; however, these conditions can hardly be expressed by numerical data. These conditions are different from country to country and from time to time. The lower population limit of urban rank may range from 200 (Denmark, Norway) to 50,000 (Japan). In Hungary, some decades ago there were only some towns with a population less than 10,000; now there are settlements of urban rank where less than 2,000 inhabitants live. Perhaps, it is not too strong an exaggeration to say that those settlements, which have been declared as urban settlements by the competent authorities, have been listed as urban ones. The percentage of urban population is rather an administrative–statistical category; consequently, its application as a parameter indicating social–economic development is misleading. Moreover, due to anti-illiteracy campaigns it is also questionable whether the percentage of illiteracy really indicates the level of education, i.e. it can be used as a parameter indicating the degree of perception. The arguments mentioned above may reveal basic conceptual problems of the model. Data representing socio-economic factors of the anthropogenic geomorphological processes concern countries; however, countries of large expanse may have an extremely varying relief and climatic features; consequently, characterization of their climate and relief conditions by one constant each can lead to sweeping generalization. Consequently, the two sides of the equation should be separated and their validity should be investigated separately (Rózsa and Novák 2008). Nir's effort to create an anthropogenic geomorphological model can be regarded as a pioneer attempt. The concept has to be refined and improved further or renewed if necessary.

References

Butzer KW (1984) Man, a geomorphological agent: an introduction to anthropic geomorphology. (Book review). Ann Assoc Am Geogr 74 (4): 647–648

Erlich PR, Erlich AH (1990) The Population Explosion. Simon and Schuster, New York

Goudie A, Viles H (2003) The Earth Transformed. An Introduction to Human Impacts on the Environment. Blackwell Publishers, Oxford

Haff PK (2003) Neogeomorphology, prediction, and the anthropogenic landscapes. In Wilcock PR, Iverson RM (eds.), Prediction in Geomorphology. AGU, Geophysical Monograph Series 135: 15–26

Hooke RL (1994) On the efficacy of humans as geomorphic agents. GSA Today 4 (9): 217, 224–225

Hooke RL (1999) Spatial distribution of human geomorphic activity in the United States: comparison with rivers. Earth Sur Proc Landforms 24: 687–692

Hooke, RL (2000) On the history of humans as geomorphic agents. Geology 28 (9): 843–846

Jansen JML, Painter RB (1974) Predicting sediment yield from climate and topography. J Hydrol 21: 371–380

Livi-Bacci M (1992) A Concise History of World Population: An Introduction to Population Processes. Blackwell Publishers, Oxford

Loh J, Wackernagel M (eds.) (2004) Living Planet Report 2004. World Wilde Fund for Nature

Nir D (1983) Man, A Geomorphological Agent. An Introduction to Anthropic Geomorphology. Reidel, Dordrecht, Boston, London

Rózsa P (2007) Attempts at qualitative and quantitative assessment of human impact on the landscape. Geogr Fiz Dinam Quat 30: 233–238

Rózsa P, Novák T (2008) Mapping Anthropogenic Geomorphological Sensitivity on Global Scale. Abstracts of First Joint Congress of the IAG Working Groups Human Impact on the Landscape (HILS) & Geomorphological Hazards (IAGeomhaz). Ruhr-University, Bochum, Germany

Simmons I (1993) Environmental History. A Concise Introduction. Blackwell Publishers, Oxford

Szabó J (1993) A társadalom hatása a földfelszínre. Antropogén geomorfológia (Human impact on the Earth's surface. Anthropogenic geomorphology). In Borsy Z (ed.), Általános természetföldrajz (Physical Geography). Nemzeti Tankönyvkiadó, Budapest

Thomlinson, R (1976) Population Dynamics: Causes and Consequences of World Demographic Change. Random House, New York

Whitmore TM, Turner BL, Johnson DL, Kates RW, Gottschang TR (1990) Long term population change. In: Turner BL, Clark WC, Kates RW, Richards JF, Matthews JT, Meyer WB (eds.), The Earth as Transformed by Human Action. Cambridge University Press, Cambridge

Index

J. Szabó et al. (eds.), *Anthropogenic Geomorphology*,
DOI 10.1007/978-90-481-3058-0, © Springer Science+Business Media B.V. 2010